User-Centred Requirements Engineering

Springer-Verlag London Ltd.

Alistair Sutcliffe

User-Centred Requirements Engineering

 Springer

Alistair Sutcliffe, MA, PhD
Centre for HCI Design, Department of Computation,
UMIST, PO Box 88, Manchester M60 1QD, UK

British Library Cataloguing in Publication Data
Sutcliffe, Alistair, 1951–
  User-centred requirements engineering
  1. Software engineering   2. Systems engineering   3. Human
  engineering
  I. Title
  005.1'2
ISBN 978-1-85233-517-5   ISBN 978-1-4471-0217-5 (eBook)
DOI 10.1007/978-1-4471-0217-5

Library of Congress Cataloging-in-Publication Data
A catalog record for this book is available from the Library of Congress.

ISBN 978-1-85233-517-5

http://www.springer.co.uk

First published 2002

Typesetting: Gray Publishing, Tunbridge Wells, Kent
Printed and bound at the Athenæum Press Ltd., Gateshead, Tyne and Wear
34/3830-543210 SPIN 10839647

Preface

If you have picked up this book and are browsing the Preface, you may well be asking yourself "What makes this book different from the large number I can find on amazon.com?". Well, the answer is a blend of the academic and the practical, and views of the subject you won't get from anybody else: how psychology and linguistics influence the field of requirements engineering (RE). The title might seem to be a bit of a conundrum; after all, surely requirements come from people so all requirements should be user-centred. Sadly, that is not always so; many system disasters have been caused simply because requirements engineering was not user-centred or, worse still, was not practised at all. So this book is about putting the people back into computing, although not simply from the HCI (human–computer interaction) sense; instead, the focus is on how to understand what people want and then build appropriate computer systems.

The book is based on my own research in the area over a number of years in several projects: basic research projects funded by the European Union (EU) – NATURE (Novel Approaches To Requirements Engineering) and CREWS (Cooperative Requirements Engineering With Scenarios) – and applying that research in EU-funded ESPRIT projects – Intuitive (database front ends) and MultimediaBroker (web-based broker services). Other research contributions come from projects funded by the Engineering and Physical Sciences Research Council – DATUM (safety critical systems), CORK (socio-technical requirements analysis), ISRE (requirements analysis using virtual reality prototypes) and SIMP (requirements analysis and performance assessment in complex large-scale systems). Besides my academic life in the field, I can lay some claim to have worked at the coal-face. Before starting my academic career I acted as my own requirements analyst in systems I developed during my PhD, and in industry as an analyst programmer; I then became what was in those days the precursor of the requirements engineer: the systems analyst. I experienced requirements creep, communication problems, irreconcilable stakeholder viewpoints, and all the other hassles that reside in RE. Of course, I still practise RE in my university domain: I have to figure out students' and employers' requirements for the courses I teach, requirements for good research projects, and indeed infer to the best of my ability your requirements for this book. Not surprisingly, I am both an academic interested in the theory of requirements and a practitioner aiming to improve what I do.

I find it helps to explain the title of my books in a preface, usually because the title reflects my motivation in writing the book, and partly to open a dialogue with the reader to explain the contents. The world is not short of books on RE, so I must confess to a slightly selfish motivation for writing yet another one. This book is primarily a review of my research in RE over the last 15 years. It also, I hope, provides a review of research by others in the area, although I should note that it was not my prime intention to review the field comprehensively. The title *User-Centred Requirements Engineering* reflects my background and research interests in the people side of requirements. Requirements engineering, I contend, should be a deeply people-centric process. Requirements, after all, originate with people, and the process of discovering them involves conversations between people; how we record and understand them involves the psychology of comprehension, and requirements for computer systems are only a small part of designing complex systems that involve people.

A sub-text for the book is the tension between theory and practice. In spite of RE's practical orientation, I argue that it does have a theoretical background and is beginning to make some theoretical contributions in its own right. The background theory comes from psychology and sociology. I cover the former in reasonable depth and give sociology some attention, but not as much as I would have liked – this will have to wait for a longer tome. I justify including psychology because RE is quintessentially about human problems of communication and understanding. Explaining how we reason and comprehend the world, as well as communicate with each other should, I hope, make for better mutual awareness when requirements engineers meet users in the miasma of uncertainty; it may lead researchers and practitioners to develop better methods and tools. RE as a field of endeavour has emerged from computer science; however, similar concerns are present in many areas of design. Within computer science, researchers in human computer interaction, information systems, and knowledge engineering wrestle with similar issues. Accordingly, I have tried to weave their contributions into the picture where I can.

To give you a reading guide of how the book is structured and my intentions in each chapter, let us start predictably at the beginning. Chapter 1 introduces the field and describes a framework for RE research and practice to date. Chapter 2 explains the psychological background behind RE problems and acts as a simple tutorial for perceptual and cognitive issues that are relevant to RE in memory, problem solving, communication, and human error. My intention in this chapter is to provide some understanding about why RE is difficult and how human misunderstanding can cause the problems we observe in getting requirements right. Any psychologists amongst you can give this chapter a miss, unless you want to look at the implications for requirements. Chapter 3 is an expanded version of a framework/survey paper I published some time ago in the *Requirements Engineering* journal. It takes the reader through the activities (or tasks) in RE, pointing out techniques, guidelines and research findings. This chapter also investigates how RE activity is influenced by the domain and is a starting point for developments such as product procurement, requirements for legacy systems, and so on. The final part of the chapter examines how reuse and different types of product influence the process. Chapter 3 is thus both a review and a framework with some embedded practical advice.

Chapter 4 covers one of the key problems in RE: how we communicate. This chapter is partly background information on psycholinguistics, but to save you from the unleavened bread of discourse theory I have included practical advice based on my research into patterns of conversation in requirements meetings. Hence the chapter gives some theory to explain how we manage to misunderstand one another, and practical patterns or templates for managing the human communication in requirements analysis. Chapter 5 continues the communication thread by investigating how we represent requirements during meetings. It consists of a review of research and a framework for thinking about representations, with practical advice on selecting the optimal set of presentations for a particular RE setting. The chapter deals with the familiar forms of requirements – lists and diagrams – but then considers how multimedia should fit into the process. This develops into more practical advice on choosing representations for the major activities in the RE process. The final part of the chapter deals with devices and technologies that we use for communication, and how computers can help group working and collaboration in requirements analysis.

Chapter 6 shifts gear wholly into the practical domain. A practical method that I have researched, taught, applied in industry and published is described. As the title indicates, it is a scenario-based approach that synthesizes much of the advice given in previous chapters into a step-by-step method. The method falls within the prototyping, requirements-by-working-with-examples tradition, so it may not fit all applications; however, it has proven its worth in several projects. Chapter 7 addresses a special concern of safety critical requirements analysis based on research I have published over a number of years. Again, a scenario-based approach is taken but this chapter also describes tools for analyzing requirements to defend against possible human error and other failures that may arise in safety critical systems. The final chapter does the futurology bit for RE and investigates how reuse and application generator technologies may improve RE. I return to the communication question and examine the prospects for the holy grail of RE: a natural language-understanding machine that automatically creates designs from our wishes. This may be fantasy, or a prospect worth more research. The chapter ends with reflections on how RE fits within other engineering and design disciplines, and how evolutionary computing might influence the future.

In any book it is a pleasure to acknowledge the help and stimulus over the years of one's colleagues, whose contributions appear in some form on the pages. I will also record the traditional caveat that should I have misrepresented their views the fault is mine. Colleagues include many on the EU basic research projects NATURE and later CREWS, notably Matthias Jarke who was always a fount of new ideas; Collette Rolland whose conceptual modelling and ideas on RE processes found their way into Chapters 3 and 4; Janis Bubenko who influenced my goal-modelling research; Klaus Pohl whose work on standard, multimedia representations and the requirements dimension have found their way into several chapters; and Neil Maiden who collaborated on the scenario-based methods and tools that appear in Chapters 6 and 7. John Mylopoulos and Eric Yu's work on conceptual modelling and their i* language appears in several places and their approach has led to many useful directions in my own work. Colin Potts started the scenario-based approach with his obstacles work and later SCENIC method that stimulated many ideas in SCRAM, while Jack Carroll's work on scenario-based design and my collaboration with him on claims appear in Chapters 5 and 6. It is also a pleasure to acknowledge the influence of some of my more formal colleagues (formal in the methods sense, not in demeanour): Michael Jackson whose research on the dependencies between the real world and the design machine has initiated a potential theory of RE; and Chris Johnson who has worked on safety critical systems and the intersection of formal and informal methods.

A substantial part of this book was written while I was on sabbatical at the University of Oregon, so my thanks go to Steve Fickas for many conversations about RE issues, systems architectures and requirements languages, and for his patience in answering other topics I pestered him with. I should also like to thank colleagues in UMIST, and Nick Gold in particular for comments on the whole text; and colleagues in the ISRE conferences and IFIP working group 2.9 whom I have not mentioned above. Finally, my thanks to my personal editor-in-chief, Gillian Martin, whose patience, attention to detail and copious unpaid assistance made writing this book a feasible task.

In conclusion, I hope you find this book interesting and useful. If having read this far you have decided you really want a more practically oriented tome, I can recommend an excellent book by Robertson and Robertson (1999) on their Volere method; the classic on requirements elicitation by Gause and Weinberg (1989) is still worth every penny; while more solid survey treatment can be found in Ian Somerville's books with Gerald Kotonya (1998) for a practical orientation, and with Peter Sawyer (1997) for a more academic view. For the more formal modellers among you, Michael Jackson's books (1983, 1995) are always stimulating. If, however, you are looking for a different perspective and I have aroused your interest in matters psychological, then read on.

Alistair Sutcliffe
UMIST, November 2001

Contents

1 Introduction

Requirements are a ubiquitous part of our lives, so it may seem strange that they have been singled out for study in computer science. Requirements and communication are inextricably intertwined. We start making our requirements clear soon after we are born by crying, usually for food. Parents soon become expert requirements engineers in inferring what their children want; but even in the cradle the dilemma of requirements is exposed. A baby's cry is ambiguous. Does he or she want food, warmth or a cuddle? How do we translate our interpretation into the right food, degree of warmth, or appropriate rocking motion?

The requirements–communication problem stays with us throughout our lives as we struggle to make our needs known to others. In professional life, requirements are more closely bound to design. Much human endeavour is directed towards creating things, be they goods or services. To create, you might be lucky and have clairvoyant inspiration; for most of us, it is better to start with a more prosaic definition of what the customer wants. All branches of engineering and manufacture have experienced the requirements problem since the industrial revolution. Creation, design and requirements have become competitive siblings. In some cases, design and creativity led requirements to the discovery of new possibilities, such as when Abraham Darby built the first iron bridge across the River Severn in 1796. Later requirements for cheaper bridges drove the designer of the Tay railway bridge to design an unsafe structure that led to the disastrous collapse of the bridge as a train crossed it during a fierce storm. So if requirements have always been with us, why has computer science discovered the need to make a special study of this phenomenon? The answer lies in complexity and the nature of software. Computer systems apply to nearly every walk of life. They appear in domains ranging from engineering to business, education and leisure. That gives computer scientists a particular problem. They have to become knowledgeable in many different domains that are not their immediate areas of expertise; hence, requirements is a special problem because of the ubiquitous nature of software. Requirements engineering is the area of computer science that addresses this concern, although it cannot claim sole ownership. Systems engineering, a product of other design disciplines, also encounters the requirements problem. I shall return to the wider perspective in the final chapter.

1.1 Motivation for Requirements Engineering

There are many definitions of requirements engineering (RE); however, they all share the idea that *requirements* involves finding out what people want from a computer system, and understanding what their needs mean in terms of design. RE is closely related to software engineering, which focuses more on the process of designing the system that users want. Perhaps the most concise summary comes from Barry Boehm: requirements are "designing the right thing" as opposed to software engineering, which is "designing the thing right" (Boehm, 1981).

Requirements have always been with us in any human act of design; for instance, the *Titanic* disaster can be construed as poor requirements analysis, in that the shipbuilder and ship owner failed to set out the precise specification of what "unsinkability" meant. The *Titanic* was correctly built to the requirements that specified that the hull be subdivided into watertight compartments, with electrically operated doors between the bulkheads that could be closed from the bridge. The owners and shipbuilders thought this equated to unsinkability, because up to three contiguous compartments could be flooded and the ship would not sink. Unfortunately they never thought that an accident would flood four contiguous compartments, which is what happened as a consequence of the glancing blow from the iceberg collision. The rest is history. The tragedy also uncovered further mistaken requirements such as the false assumption that lifeboats sufficient for all the passengers and crew would not be necessary. This assumption saved money during construction, but with terrible consequences.

We have not improved our practice much since 1910, and mistakes in RE still cause problems. Modern aircraft are controlled by computers. An A320 airbus had just taken off from Gatwick airport when the pilot discovered the aircraft making an unexpected turn to the left. The flight crew struggled to control the aircraft and thought that the aerilons (flight controls on the wings, which alter the angle of banking as an aircraft turns) were not working correctly, over-compensating for slight changes in direction. They eventually managed to land the aircraft correctly having struggled with the emergency flight management procedures. These proved to be impossible to find via the menu hierarchy in their cockpit VDU displays but the pilot, luckily, had an old paper manual. The safe return of the aircraft was due to the skill of the flight crew rather than the system design, which did not diagnose the problem in the first place and which then hindered their attempts to correct the failure.

The cause of the near accident was that a maintenance crew had been servicing the hydraulics that control the wing slats. The wing slats automatically open and close to equalize pressure on each wing when the aircraft turns. The hydraulics can be disabled by placing them into service mode, so that damage by manual manipulation of the slats is avoided. When the hydraulics are disabled the slats open and close with small changes in air pressure leading to instability. Unfortunately, the maintenance crew left the hydraulics in maintenance mode so the slats did not work on take-off, and nobody noticed beforehand. The maintenance crew were new, poorly trained and under time pressure to complete a service on schedule. The management culture of pressure and inadequate resources led to the window of opportunity for disaster. The designer had never thought that a requirement for a warning light showing that the hydraulics were in maintenance mode would ever be needed; neither did the pre-flight checklist consider this to be a hazard worth checking.

Computer systems have made the requirements problem worse because we build systems in many different domains. The requirements engineer (or designer) has to understand not only what the user wants, but also the implications of the domain and what is achievable. Since gathering requirements inevitably involves communication between people, and natural language is prone to misinterpretation, requirements analysis has been a frequent cause of system failure.

The failure to undertake a thorough requirements analysis is illustrated by many well-publicized computer system disaster stories. One of the most prominent in recent years was the London Ambulance Service's Computer Aided Dispatch System (Finkelstein and Dowell, 1996; HMSO, 1993). This was intended to replace the manual system of answering emergency telephone calls from members of the public, finding out where the emergency was and then dispatching one of the available ambulances to the location of the accident. Superficially this was a simple resource allocation problem. While nearly every possible mistake in software development and project management was made in this case, poor RE played its part. The software developer was a small systems house with little experience in the domain of ambulance call-dispatch systems. The systems analysts, nominally in charge of requirements, were the in-house software development team of the London Regional Health Authority. The users were the operators in the call-dispatch centres whose job it was to allocate ambulances to emergency calls, and the ambulance crews themselves. Little, if any, requirements analysis was carried out with these real users. The Regional Health Authority development team specified the requirements. There was no consultation or opportunity for feedback until the system was ready for deployment. At that stage, performance problems were so bad that just getting the system to work was an uphill struggle. Poor requirements analysis failed to detect several problems: radio blackspots where the ambulance crews could not be contacted, poor user interfaces on the mobile data terminals which resulted in the ambulance crews not reporting call progress accurately, with the knock-on effect that the system database became inaccurate, causing the automatic call allocation program to lose ambulances, dispatch ambulance crews who were not free, send several ambulances to the same call, and so on.

The London Ambulance Service is not alone in the litany of RE failures. Government departments and the armed services in particular have a long track record of getting it wrong. The UK social services tried to computerize the claims payment systems in the 1980s, but had to cancel the whole development because the requirements could never be stabilized. The updated UK air traffic control system was supposed to have been implemented in 1998 but at the time of writing has yet to go live, partly because the requirements for the complex 3-D visual displays took so long to finalize. A similar problem has emerged with Eurocontrol (Europe-wide air traffic control) where requirements for flight co-ordination have caused endless problems. In the USA the Pentagon's systems have also had their share of requirements disasters, ranging from Patriot missile radars that could not see a Scud warhead among missile debris (missed requirement that Scud missiles were so badly engineered they fell apart on the way down) to logistics systems that were never implemented (Potts, 1999).

So the penalty of getting RE wrong is high. Systems may fail, but even if they do not, their use is sub-optimal or design costs are wasted. Studies of banking systems showed that users only used 30 per cent of the functions provided (Eason, 1988). There is plenty of evidence that RE is a difficult task. Stories of spectacular system failures (e.g. the London Stock Exchange Taurus system and the London Ambulance call-dispatch system outlined above) point to poor requirements capture or even failure to make any serious attempt to capture user requirements. So is it just a problem of motivation and education? In spite of several well-publicized disasters and an invisible legacy of system failures, many systems have been developed successfully, and most companies now depend on computerized accounts, payroll, and stock control systems. The key to the problem is change. If organizations kept to the same practices and the world stayed the same, then requirements would be easier; but they don't. Managers want to improve business practices, governments change tax laws, businesses have to respond to challenges from competitors. Consequently, requirements can never be said to be complete and computer system designers are shooting at a moving target. This problem has been elaborated by Lehman and Ramil (2000), who point out that most software has to be designed to evolve.

The motivation for RE is simple: to reduce the high cost of misunderstanding between users and designers, so that computer systems are built to do what the users want, on time and at a reasonable cost. The high cost of errors incurred during many system failures can be attributed to mistakes in requirements analysis (Bell and Thayer, 1976). Although RE only amounts to 10–15 per cent of the overall cost of system development, the consequences of getting requirements wrong can have a disproportionately high impact. The longer a mistake remains undetected during development the more expensive it is to fix. Changing a requirement just takes a bit of word-processing, changing a design involves undoing the specifications and checking ramifications of new requirements, but changing code means throwing away programmers' time and possibly destabilizing a design. The economic arguments for RE are summarized in Figure 1.1.

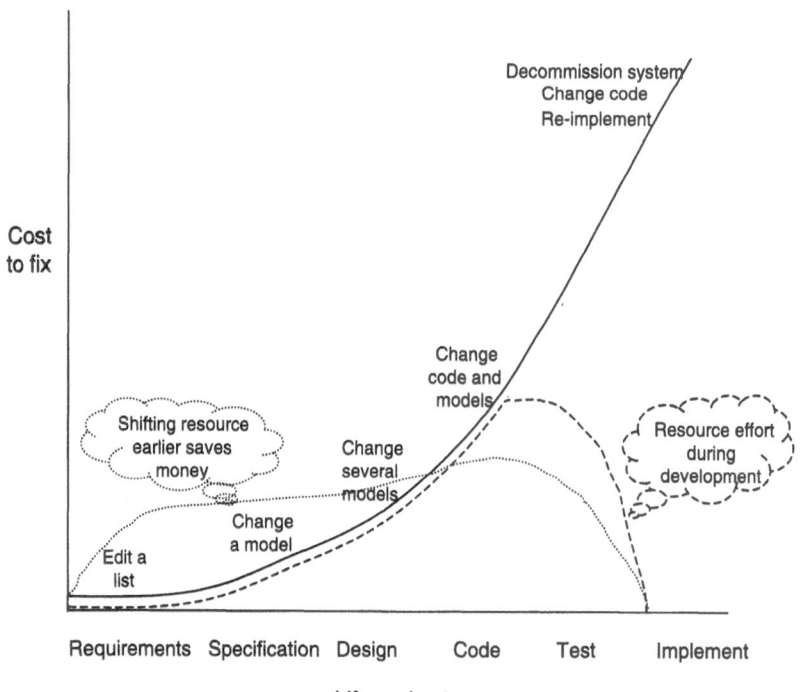

Fig. 1.1 Costs of change and the development cycle.

Investing in RE early will save money later on, but it can be difficult to decide when requirements analysis should stop. First there is the "fickle user problem": people tend to discover new requirements just when the analyst thought everything they wanted had been captured. Secondly, requirements analysis is difficult because of the "moving world problem". All the analyst can do is to get the best possible picture of the world when capturing requirements, try to anticipate the future, and then design software so that it is flexible and can adapt to change. While methods and technology can help with requirements analysis and flexible design, nobody possesses 20/20 foresight, so anticipating the future is a black art. As we shall see later, however, there are ways in which designers can frame their ideas about the future.

1.2 A Little History

Before RE, requirements were subsumed in systems analysis. This area produced structured development methods such as SA/SD (structured analysis/structured design: De Marco, 1978); SADT (systems analysis and design technique: Ross and Schoman, 1977) and SSADM (structured systems analysis and design method: Downs, Clare and Coe, 1992). These methods all referred in passing to requirements analysis and generally started with a divide-and-conquer approach of carving a system into successively smaller pieces and then defining the functions or goals that each part of the system was supposed to achieve.

So why was systems analysis not good enough and a whole new sub-discipline of RE needed? Part of the answer is rooted in academic communities. Systems analysis belonged to information systems, a community concerned with methodology and conceptual modelling. It is probably true to say, however, that the concept of requirements analysis in information systems was suffering from an intellectual strait-jacket of top-down functional decomposition. Requirements were captured and listed as users' goals, then elaborated as a set of functions represented in data flow diagrams, SADT processes or whatever the analyst's favourite conceptual modelling language was. The "structured methods" approach to requirements analysis was challenged in the 1980s with JAD/RAD (joint/rapid applications development) techniques (DSDM, 1995) which tried to overcome the cumbersome and time-consuming process of modelling with the use of workshops, scenarios and brainstorming sessions to encourage more user involvement in design. The "top-down" analysis approach of structured methods was also challenged by the object-oriented community who proposed modelling the domain rather than establishing goals/functions.

This led to the current generation of structured object-oriented methods: object-oriented systems engineering (Jacobson *et al.*, 1992); the object modelling technique (Rumbaugh, 1991); and object oriented analysis/design (Coad and Yourdon, 1991), to cite but a few of the contenders. More recently the object-oriented community created a *de facto* development standard as UML (unified modelling language) and the unified process (Rational Corporation, 1999), but this has little to say about requirements analysis apart from advocating use cases and scenarios, which will be explained later.

The other influence on RE was software engineering, a community who focused on formal specification and delivery of reliable software products, often for real time telecommunications and safety critical applications. Software engineers found that in spite of years of formal specification many systems were not accepted by users or failed in circumstances that had not been anticipated. RE has grown from, but also contributes to, all these areas. It is related to research in software engineering and conceptual modelling by sharing models and formal languages drawn from that research. RE is also concerned with process, although not always in the structured methods sense. RE research, however, grew from the need for techniques and tools that complement software engineering methods, so RE can be seen as a collection of techniques that are recruited to a more general process according to the users' and designers' needs.

The foundations of RE were set out in a collection of papers (Thayer and Dorfman, 1990) and special issues of *Transactions on Software Engineering* (IEEE-TSE, 1991, 1992). These were followed by IEEE symposia and conferences (Finkelstein and Fickas, 1993; Davies and Hsai, 1994). These events revealed a diversity of research issues and industrial practice that can be loosely associated with defining what to build rather than how. Lubars, Potts and Ritcher (1993), in one of the few investigations of RE practice in industry, reported that ambiguity and changing requirements were a constant problem and that developers preferred organizational to technological solutions for RE problems. More recently, field studies of system development practice (El Emam and Madhavji, 1995) have indicated that changing requirements, lack of trained manpower and inadequate methods are responsible for system failures. RE research has tended to be dominated by large customer-driven systems, typically in the defence sector. In these contexts, requirements are complex and often driven by a super designer's vision, whereas market-driven requirements arise out of a more creative brainstorming approach (Lubars *et al.*, 1993).

1.3 People, Communication and Requirements

Another reason why requirements analysis is hard is because of people. We all have our own ideas and viewpoints. Many are unconscious attitudes that we only become aware of when someone challenges our beliefs. Furthermore, we keep some facts and attitudes to ourselves for a variety of private and political reasons. The upshot of this is that getting a complete set of unadulterated facts from people is unlikely even with a set of co-operative, honest and well-motivated users. The practice of requirements analysis thus has to combat problems of human communication that involve psychology and sociology, such as tacit knowledge, the ambiguity of natural language, and the role of power and personality that influence attitudes and opinions.

● *Tacit knowledge* – as we learn any skill or become familiar with a domain, we no longer think about it in conscious terms. Consider driving. Unless you are a learner driver you don't think about changing gear or steering the car; these are automatic skills acquired years ago. Similarly, you don't think about your route to work; you just know it. So when someone comes to ask you about your experiences in getting to work, you are unlikely to tell them about the route in detail or how you drove your car. The same problem exists in all domains. Users are experts, so they don't tell the requirements analyst about the obvious facts, yet these are often important in understanding how that domain works. And this doesn't take into consideration the facts that people *want* to hide.

- *Ambiguity* – even if we try to report what we do, we frequently do so inaccurately. English and most other natural languages are wonderfully flexible means of expressi ıg ideas, but the disadvantage is that expression is not always precise. A classic exa..nple is conjunction in logical procedures; when we say "and" we sometimes mean "both", sometimes "either". For instance, "I go to the bank and building society to get cash" – does this mean I always go to both the bank and the building society, or sometimes to just one of them? Ambiguity is inherent in logical operators, comparisons (greater, equal to, etc.), as well as imprecise expressions of procedures. Sometimes this is a consequence of poor expression, but it may also be a symptom of vague thinking, as we discover when following street directions from a well-meaning but uninformed local inhabitant.

- *Attitudes and opinions* – people tend to pick up attitudes and opinions from their family, peers and news media. When asked to justify an objective or goal, they may give an opinion that has never been thought through. Requirements analysts therefore have to challenge attitudes to find out which ones have a rational justification and which are a matter of folklore. When challenged, people may get into trench warfare based on dogma rather than arguments based on their merits. Furthermore, opinions are often influenced by internal politics and the power structure of organizations. Some employees, conscious of job insecurity or an authoritarian boss, may give false opinions to placate a "party line". Social factors, such as power and responsibility, influence the expression of requirements (Goguen, 1993). People often fear change itself. Requirements analysis exposes problems and hidden agendas in organizations. Fact gathering has to tackle problems of public versus private versions of truth which users may, or may not, wish to communicate (Harker, Eason and Dobson, 1993; Maiden and Rugg, 1994). Eliciting true attitudes is one of the most difficult tasks the requirements engineer has to tackle. Failure to detect political problems can lead to systems being implemented without the consent of all the stakeholders, or to problems remaining hidden until it is too late. Early detection of false attitudes alerts the analyst to potential conflicts and the political danger of getting trapped on one side or another. Reconciling viewpoints to get a consensus view is another difficulty. The analyst can find him or herself in the middle of an inter-departmental war, for example between marketing and customer services who have very different high-level goals and do not understand each other's point of view.

1.4 A Framework for RE

Several definitions of RE have been given. For instance in terms of the outcome, "a requirements specification should tell a designer everything he needs to know to satisfy all the stakeholders – but nothing more" (Sommerville, 1989). Alternatively, the principal RE issue described by Bubenko (1993) is "how to proceed from informal, fuzzy individual statements of requirements to a formal specification that is understood and agreed by all stakeholders". Dubois, Hagelstein and Rifaut (1989) used the term to refer to the part of the development life cycle in which the needs and requirements for the user community are investigated and then abstracted to create formal specifications. Zave (1995) proposed a more elaborate taxonomy of problems in requirements engineering that helps to scope the field. Her definition of RE was "the branch of software engineering concerned with real world goals for, functions of, and constraints on software systems. It is also concerned with the relationship of these factors to precise specifications of software behaviours, and to their evolution over time and across software families". At the top level Zave divided RE in two dimensions. The first is the *problems* of requirements engineering, decomposed into: specifying system behaviour; problems of investigating the goals, functions and constraints of software systems; and problems of managing the evolution of systems and families of systems. The second dimension is RE *solutions*: the type of research in the field, such as state of practice reports, process-oriented solutions (methods), product-oriented (tools), case study application of solutions in the real world, evaluations of approaches and measurement of success. One of the tensions these definitions reveal is the focus of RE on software or on requirements for large-scale systems (including people), and how reuse fits into the picture.

Trying to state the objectives (or requirements) of RE reveals some of the dilemmas that need to be solved. For instance, the following objectives for RE may conflict:

● to capture a complete set of requirements from users;
● to analyze the users' requirements accurately and understand all the implications inherent in those requirements;
● to specify how those requirements should be met in a design;
● to complete requirements analysis within acceptable constraints of time and cost.

Given patience and infinite resources it may be theoretically possible to get a set of near-perfect requirements; however, within the resource and time constraints of the real world this is unlikely. The adjective "near-perfect" is used because requirements analysis faces another dilemma: the world keeps changing. When the analyst has captured a complete set of requirements they inevitably refer to yesterday; meanwhile the world has altered, and users will have changed their minds. Requirements, therefore, are at best a compromise.

The three dimensions proposed by Pohl (1993) (see Figure 1.2) illustrate three major problems that RE has to solve: namely modelling the future application in a

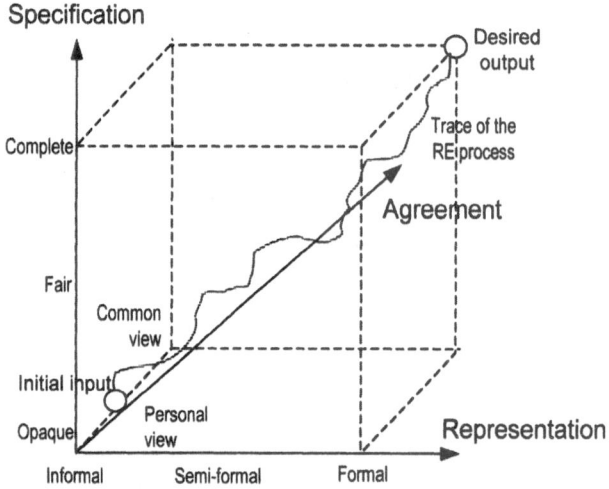

Fig. 1.2 The three dimensions of RE.

more complete manner, modelling with more formality, and with all the stakehold-ers agreeing requirements. Input to the process starts with coarse-grained and ambiguous statements about the users' requirements for the intended system. Users may have different visions of requirements or only partial and incomplete ideas about what they want. Input is characteristically informal in its representation, imprecise and personal since requirements are initially held by individuals and fre-quently conflict with one another. However, the desired output from RE is very dif-ferent. It should be a complete system specification, within the constraints of available resources, using a formal language, and agreed by all involved.

The thread in Figure 1.2 traces the emerging requirements specification, as it becomes more complete, accurate and shared. On the specification dimension, RE has to guide the discovery, refining and validation of requirements as they become more thoroughly understood and complete. Representation has to support expression of requirement statements and models in natural language, semi-formal graphical nota-tions such as data flow diagrams (DFDs) and entity relationship (ER) diagrams, and formal notations. Finally the agreement dimension has to support trade-offs between requirements and different stakeholders' views, negotiation and co-ordination.

Another view of requirements sees the problem of transforming a current system into a new, desired system. This view focuses our attention on the problem of under-standing existing systems before we can design the necessary change. The view is elaborated in the four worlds framework (Jarke *et al.*, 1993) which is useful in understanding how requirements change and how they can be partitioned into dif-ferent viewpoints (Figure 1.3).

The first world is where requirements usually start in the usage world of the user. The usage world describes requirements in their context of a domain and, possibly, an existing system. Requirements reflect the wishes and objectives of users, or their problems with a current system that needs fixing. Requirements are elicited or acquired from users in the usage world, and transferred into the subject world as models and representations of the domain. This involves abstracting a generalized picture from the detail in the usage world. Generalized pictures become models and specifications of the world and the designed system. Notice here the tension between requirements as single goals and requirements embedded as components of a model or specification. Should requirements be simple lists of functions or goals, or do we need models of objects, tasks, and processes to understand them in context? The answer is probably both, but the relative importance of lists and mod-elling is more difficult to judge. From the subject world, which represents part of modelled reality (also referred to as the universe of discourse) requirements are transformed into the system world. Here requirements become specifications of what will work in the new designed system. Requirements therefore progress from "why" questions of users' goals in the usage world to "what" questions of functional specifications in the subject world and finally "how" questions in the system world.

However, the progress from requirements to design is not as clear as this frame-work would suggest. Users frequently do not have a clear vision of what they want. If they do, their goals may be poorly formed and fuzzy. Furthermore, requirements are linked to people's knowledge of what is technically possible. This leads to the inevitable intertwining of requirements with design, to paraphrase Swartout and Balzer (1982). Only by creating some vision of the future can we be sure that it embodies our requirements and even then our understanding will be limited by how faithfully the current vision (or prototype) represents the final product. Naturally, the more users change their minds during requirements analysis, the more the original vision will depart from the final design. So at the heart of RE there is a paradox: users don't know what they want until they get it, and when they get it they see how it could be improved or they don't like it. Trying to overcome this 20/20 foresight problem lies at the heart of RE.

The final world in Figure 1.3 is the development world. This is the realm of the programmers and software developers. The development world has its own requirements such as constraints on design for interoperability, maintainability, reliability, etc. In addition, software used to develop computer systems imposes another set of requirements which the requirements engineer has to deal with – for example, can the application be delivered in Java on client–server architecture, and if so is the data environment secure?

Having looked at some of the perspectives of RE, it is now necessary to investigate how different starting points for requirements may affect the process.

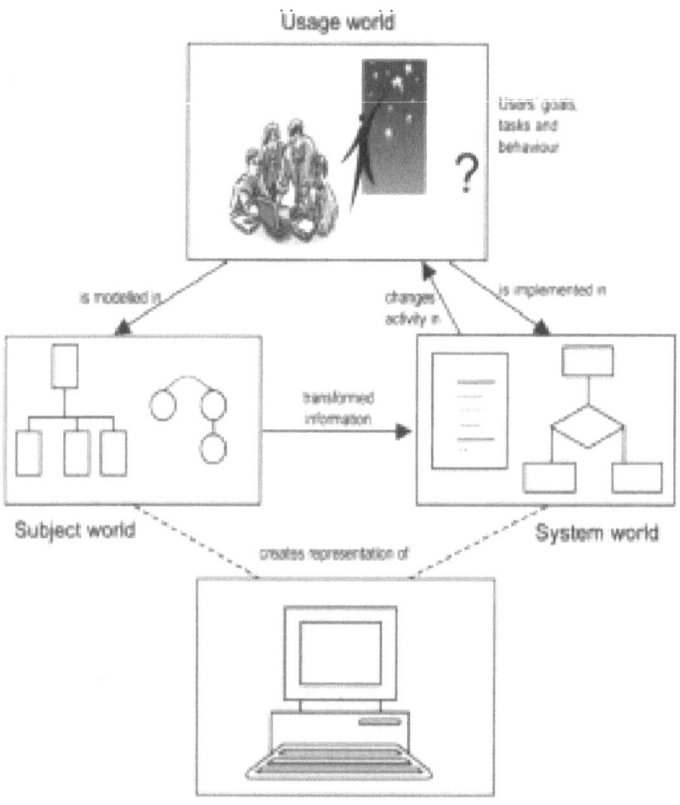

Fig. 1.3 The "four worlds" view of system development.

1.5 Requirements Types and RE Pathways

Although RE is often assumed to start with top-down decomposition to create goal hierarchies, it may also start with problems in an existing system rather than intentions to create a new system. Furthermore, user requirements may also be promoted by examples of other successful systems. Different applications affect the course of the RE process; for instance, COTS (commercial off-the-shelf) software selection is different from bespoke development; while information systems and real time domains make different demands on the RE process. Requirements come in many different shapes and sizes. Some may be high-level goals while others may be detailed rules and constraints. To give some idea of the range, the requirements in a library circulation system could include:

- the need for a complete sub-system or high-level functions: "the system will have facilities for auditing book stock so that old and redundant stock can be eliminated";
- specification of a more detailed function: "the circulation control system should calculate fines on overdue loans";
- a statement about how a function should work: "fines should be calculated as the number of days overdue (current date – due date) multiplied by a configurable fine factor";
- constraints on how the system should operate: "data on reader fines should be secure and not publicly accessible";
- statements about performance: "all search requests must be completed within 30 seconds of submission";
- implementation constraints: "the system must operate on a Linux platform as well as Windows NT".

The above requirements vary in the amount of detail provided, but even ones which look precise suffer from ambiguity. Take the third requirement in the list. This looks precise, but it hides further questions. How many configurable parameters are required, such as separate fine rates for standard readers, old age pensioners, readers who are frequently late, etc.? Many requirements are inaccurate and ambiguous. The RE process has to clarify what is meant as far as possible.

Not all requirements come from people. Requirements can be imposed on a system by laws of physics and facts of nature. For example, in a fly-by-wire avionics system the computer has to control an aircraft so that it flies at a certain speed, direction, height and orientation. Failure to do so will cause the aircraft to stall, or worse still, to become over-stressed and crash. Requirements in this case include the laws of physics such as gravity, aerodynamics and physical stress limits of the aircraft. The system must conform to these facts and process events in a specific order. This problem of requirements emanating from implications of events has been investigated by Jackson (1995) in several refinements of his method which derived entity life histories from explicit consideration of real world facts and behaviour. In more recent papers, the correspondence between real world events and system behaviour has been expressed in terms of optative requirements to which the system must respond (Jackson and Zave, 1993). These are obligations on the system, or required behaviour that it must carry out to ensure successful operation. According to Jackson's approach to RE, it is necessary to distinguish between descriptions of the world as it exists and designations that reflect assumptions about properties of designed systems. Requirements emerge from understanding how the system should behave – *indicative requirements* – and other necessary responses by the system to change the world – *optative requirements*. Requirements are statements of what machine specifications should do given assumptions about the real world domain.

Some requirements are imposed externally as constraints on design required by law, others arise from problems which need fixing, and some are performance qualities rather than functions which might be implemented in software. Roman (1985) provided the first list of issues in RE and drew attention to the need for modelling functional and non-functional requirements. There are several taxonomies of requirements (see Pohl, 1996; Mazza *et al.*, 1994), which will not be elaborated here; however, it is necessary to explore some fundamental categories. A commonly held but disputed distinction is between functional and non-functional requirements:

- *Functional requirements* are statements about what a system should do, how it should behave, what it should contain, or what components it should have. Functional requirements are initially expressed as goals, e.g. "the search facility will find books that match the user's request"; or activities, e.g. "validate order", "monitor temperature". Later, requirements become specifications expressed as entities, attributes, actions, events or states: the familiar semantics of software engineering modelling languages. Functions are elaborated processes or procedures which can be implemented as software algorithms and data structures.

- *Non-functional requirements* (NFRs) are statements of quality, performance and environment issues with which the system must conform. Some examples are reliability, maintainability, portability (properties of the design), safety, security, scalability, accuracy, usability, and performance. NFRs are qualities and performance criteria that are not directly implementable in software. They can be further sub-divided into:
  - performance criteria such as throughput volumes (system must handle 10,000 transactions per working day), reliability, response time;
  - design-related constraints such as maintainability, interoperability (e.g. system works on both Microsoft and Linux platforms), security, usability.

The distinction between functional and non-functional requirements, when examined in detail, may crumble. NFRs set standards or benchmarks by which the system will be assessed but they also have design implications. Refining NFRs involves decisions about how well a design can meet the required criteria so, rather than being implemented, NFRs are satisfied to some extent by design. Mylopoulos and his colleagues (Mylopoulos, Chung and Nixon, 1992; Yu, 1994) describe NFRs as "soft goals" that, unlike functional requirements, cannot be completely specified in terms of software; instead, these requirements are "satisfyced" to some extent by functional requirements in a design. NFRs do not directly refer to what should be designed; instead they are statements of quality or performance criteria, for instance "the system shall process 1000 orders per day, with an average time per order of 20 seconds". How they are satisfied may involve defining functional requirements, so the dividing line between the two is not sharp. An example might be a safety requirement that "the autopilot system will fly the aircraft to ensure no collisions occur". This is a bit vague so the designer might unpack what "no collisions" means as "the autopilot will detect any aircraft which comes within an envelope of 500 metres and take avoiding action". However, "avoiding action" has not been defined. This will become a set of rules or procedures about how to control the aircraft, such as if an aircraft approaches from behind then speed up, from below then climb, etc. In this manner a non-functional requirement has become refined into several functional requirements for monitoring the proximity of other aircraft and procedures for avoiding action. However, not all NFRs can be unpacked in this manner. Performance requirements become metrics or benchmarks against which the system design will be assessed.

One approach to dealing with benchmark-style NFRs (Briand *et al.*, 1995) is the goals-question-metric technique for software engineering quality assessment, even though Briand *et al.* do not explicitly refer to requirements. The essence of their method is to set goals as quality criteria that are measurable, with a metric linked to a particular need that is expressed in a question. For example:

Goal: To achieve an acceptable response time of less than 0.02 seconds for user system interaction.

Questions: How quickly will the system respond to users' editing operations with a normal workload?

Metric: Delay from the end of a key press to the appearance of a visible change on the screen.

Hence NFRs will become refined into functional requirements as well as setting a necessary and sufficient criterion by which the system can be judged. For example, safety as an NFR will have design implications for how safety is delivered as well as being unpacked as a metric to specify what an acceptable level of safety means: in air traffic control "the system will prevent mid-air collisions, and the near-miss rate, where two aircraft pass within 1 mile or 500 vertical feet separation of each other, will be less than 1 per 100,000 aircraft movements".

1.6 Constraints on Design

Another way of thinking about requirements is to classify users' needs into goals or functions, attributes which describe the properties of qualities of the desired system and constraints that define the limitations on design. To elaborate these categories:

● *Goals* are functional and non-functional requirements that describe what the users want the system to do.

● *Attributes* are qualities or properties of the desired system, which may be functional or non-functional in nature, e.g. the system will be safe, reliable, operate efficiently, inexpensive, have aesthetic appeal.

● *Constraints* are conditions or laws that the system will have to obey during operation or during design, e.g. the system should operate in a temperature range of 0–30ºC, and fit within a space of 1 cubic metre. Constraints fall into four sub-classes:

  – Constraints on the physical shape and size of the product. In software these may be megabytes of memory or disc storage space; more generally they are limits on the size and shape of the product.

- Environmental constraints that set out the expected means and ranges of operating conditions, e.g. temperature, pressure, vibration, different locations, dust, dirt, noise, etc. While software is shielded from most of these environmental factors, human operators and other equipment are not, hence these constraints are important.
- Cost. This is always a key limitation on the designer's (and users') ambition.
- Legal constraints that set non-functional requirements for safety, reliability, usability, security and privacy.

Ideally, requirements would be developed for systems with an infinite budget and no dependencies on existing systems. Unfortunately such green-field applications rarely exist, and budgets are never unlimited. Requirements have to be specified that are realistic and achievable within the costs and constraints of an organization. Constraints vary from legislation imposed by government that may have implications for health, safety or working conditions, laws that influence the applications itself (e.g. tax laws and payroll calculations), to constraints on processes and procedures imposed by company policy (e.g. not offering discounts to poorly paying customers). Increasingly, requirements also have to be specified for system upgrades or for new systems that have to co-exist with previous implementations. Legacy software consists of suites of old programs, usually written in COBOL, that most companies possess to run their basic transaction processing applications, i.e. sales order processing, inventory control, general ledger accounting, etc. Legacy systems are often critical to a company's survival, so new systems have to be built as front-ends to the older systems. New requirements may imply changes to legacy systems which, given their antiquity, are poorly designed and hard to change. Since requirements were rarely documented in older systems, deciding requirements for modification or how new systems should fit within a legacy system can be a challenge.

Requirements can be reverse engineered or recovered from legacy code but the process is difficult. Programmers are notoriously bad documentors so the analyst is often left with just the code itself. The research area of reverse engineering (for overviews see Layzell, Freeman and Benedusi, 1995) deals with this problem of restructuring old code to make it more maintainable and reliable. If the code is well-designed with a modular structure then there is some chance of success that the original designer's intentions might be discernible; however, most old code has as much structure as a plate of spaghetti, so reverse engineering requirements can be a forlorn task. The users' goals were rarely documented in older systems and guessing them from algorithms and data structures is a nearly impossible task.

Cost is one of the major constraints on any project. Customers and users, if given the choice, will create requirements wish lists that could consume many times the development budget, so the requirements analyst has to prioritize requirements in collaboration with the users. Processes for doing so will be covered in Chapter 4. Some applications may not necessitate development of new code; instead the solution is to purchase COTS software. In this case the constraints are the available products and the range of functionality they provide. The requirements process therefore becomes a goodness-of-fit optimization, matching customer requirements and products' properties within the constraints of costs, delivery time, maintenance support, etc. A variation on COTS is configurable products, such as Enterprise Resource Plans (ERPs) from vendors such as SAP and Peoplesoft. These products provide generalized business application packages such as personnel, payroll, logistics, sales order processing, etc. Such packages can be tailored to the customer's requirements within certain limits. They impose fewer constraints than COTS products but more than in a bespoke development of new software. For ERP products the requirements process has two phases: first, selecting the ERP vendor and application package; then, requirements for tailoring the package to the customer's business. The first phase follows a COTS-like goodness-of-fit process, while the second phase is constrained by the customization facilities provided (see Sutcliffe, 2002).

In summary, the requirements process is constrained by many realities of the environment in which it takes place: time, costs, laws, standards, and available resources. The transformation of requirements into design is also constrained by possibly having to co-exist with legacy systems, and standards for interoperability on operating systems, networks, etc. Finally, the requirements process might be limited by what is available in the market place for COTS-style developments. The art of successful RE is achieving the optimal within the bounds of the possible.

1.7 Documenting Requirements

Requirements are usually written down as natural language statements, but natural language is prone to misinterpretation. One response has been to advocate use of formal, mathematically based specification languages which make meaning more explicit. Unfortunately, formal specifications are not comprehensible to most users, so formal languages present a communication barrier. There is no easy escape from the dilemma of easy to understand but ambiguous natural language versus formal but inaccessible specification languages (see Chapter 3). Requirements need to express a range of possibilities in general terms during the early stage of the process, while being more precise later on (Fickas and Feather, 1995).

To avoid documents containing long, dense texts that nobody wants to read, requirements are formatted into structured lists to help understanding and retrieval. Structured requirements documents help by classifying and indexing requirements into categories so they can be found quickly and traced. Each requirement is still expressed in natural language but the statement is short and terse to reduce ambiguity. Structured requirements documents such as the standard defined by IEEE 830 (Mazza *et al.*, 1994) recommend the following headings:

1. Introduction
    1.1   Purpose of the requirements documents
    1.2   Scope of the product
    1.3   Definitions of acronyms and abbreviations
    1.4   References
    1.5   Overview of remainder of the document
2. General description
    2.1   Product perspective
    2.2   Product function
    2.3   User characteristics
    2.4   General constraints
    2.5   Assumptions and dependencies

3. Specific requirements

(including functional, non-functional and user interface requirements, performance benchmarks, design and database constraints, system attributes and quality
characteristics)

4. Appendices

5. Index

Requirements document management tools (e.g. RequisitePro, DOORS, CRADLE)
provide configurable classification structures, with defaults based on the IEEE standard or similar layouts. Requirements as statements and lists need to be accompanied by diagrams and other information that helps their interpretation. Some
authors refer to requirements documents as requirements specifications, by which
they mean requirements statements accompanied by more formal models that can
make the interpretation of natural language more precise. A further means of
recording requirements to reduce ambiguity is to use templates. These provide a
form that links the requirements with associated facts such as who authored (or
captured) the requirements, from whom (the user), when, etc. A good example of
requirements documentation templates (Figure 1.4) is given in the Volere method
(Robertson and Robertson, 1999).

Requirements may represent the views of several different groups of people,
called "stakeholders". For example, requirements documentation may be created by
the following stakeholders:

● *Customers* who will be purchasing the system or otherwise financing its development. Customers are interested in making sure requirements fit with their objectives to deliver a business or organizational benefit.

Requirement #:	Requirement Type:	Event/use case #:

Description:

Rationale:

Source:

Fit Criterion:

Customer Satisfaction:	Customer Dissatisfaction:

Dependencies:	Conflicts:

Supporting Materials:

History:

Fig. 1.4 Requirements documentation from the Volere method.

- *Users* – people who will actually operate the system. User stakeholders are concerned about more detailed requirements of system functionality and usability.

- *Managers* – stakeholders who are not direct users but manage the system. Managers may be the same as customers in some systems or may represent the department that is receiving the new system. They are interested in requirements as system outcomes to achieve business objectives.

- *Software engineers* – designers who have to translate requirements into software designs and code. They need requirements to be as precise as possible.

- *System testers* and quality assurance – stakeholders who will be responsible for ensuring that the designed system is reliable and meets with performance criteria. These stakeholders are interested in non-functional requirements and constraints.

- *System maintainers* – people who have to keep the system running when it is in operation and modify it. Maintenance personnel are rarely consulted in requirements analysis yet they depend on accurate requirements documentation more than most.

Different stakeholders have different views on a requirements document. Some want detail, others need to check that requirements will fulfil their objectives. In the early stages of RE there is considerable merit is having incomplete and ambiguous requirements. Trying to nail everything down in detail too early may constrain exploration of the problem and cause people to commit too early to a poorly thought out solution.

As RE is primarily a computer science discipline there is a hidden assumption that all requirements are about software. This is not true. Many requirements will be satisfied by management decisions to allocate resources or change human operating procedures. A useful distinction is between system and software requirements:

- *System requirements* – requirements in the wider sense that record any particular need for a new system. These requirements may specify obligations for human operators, organizational resources, or machinery, as well as implying software requirements.

- *Software requirements* – requirements in a narrower sense that pertain to intended software functions. Software requirements are derived from system requirements and make assumptions about design, i.e. what is going to be automated in software.

In conclusion, requirements documents are formatted lists of natural language statements accompanied by additional text, sketches and diagrams. These documents are used by several different groups of stakeholders who have different interests and focuses when they read requirements documents. Interests can be served by a continuum of representation from information statements in natural language to more precise specifications, but finding a single lingua franca that can be both precise and sufficiently flexible to meet the needs of all stakeholders is a difficult task. Chapter 4 explores how combinations of representations can address this problem.

1.8 Summary

RE has developed to tackle the vexed problem of obtaining the needs for a new system as accurately and completely as possible. Failure to carry out effective requirements analysis has led to many system disasters, and requirements errors become progressively more expensive to cure as system development progresses. The subject area of RE has evolved from information systems, object-oriented development methods and software engineering. RE is difficult because it has to deal with requirements which may never be complete because users discover requirements only when they experience a design and because the external world changes and creates new requirements after a system has been implemented. Requirements analysis has to deal with communication problems and the ambiguity of natural language. People may not provide correct information because their knowledge is tacit, or for political and personal motivations to hide sensitive facts. Requirements can be divided into functional and non-functional requirements. Functional requirements are captured as user goals and refined into specifications for the design. NFRs become benchmarks or performance and quality criteria that a design has to achieve, but during analysis functional requirements are discovered to achieve them. Requirements are documented in structured lists defined in standards and using templates that record the authors, stakeholders and associated information. Different stakeholders will have differing needs and views on requirements documents.

2 Understanding People

2.1 Introduction

To understand the requirements process we need some knowledge about our own processes of understanding problems and communicating with people. Requirements analysis is a cognitive process in which we understand problems, learn about domains and negotiate to achieve what we want. We have to understand details describing the real world and then abstract these to create models of the domain. The process of abstraction, in cognitive terms, involves perception and comprehension (making sense of what we hear and see) and learning. Requirements analysis can be seen as discovery-based learning. We start with a vague understanding of what is required and of the domain in which the new system is to be designed. By gathering information and requirements we gradually learn about the goals the new system should fulfil. Requirements is also a form of problem solving. Once the desired state of the world has been understood, i.e. the requirements for a new system, the designer has to solve the problem of achieving them. The inevitable intertwining of requirements and design is a problem-solving process in which the designer uses memory of previous successful solutions, software engineering methods and creative thought to achieve a new design.

Given that RE involves cognition in several ways, understanding the psychology of problem solving, memory and human communication can help us to interpret the recommendations of RE methods and techniques which academics and practitioners have advocated. More importantly, background knowledge allows us to reason from first principles about why a method or model might be suitable for a particular activity. This chapter describes the background in cognitive and social science that impinges on RE.

2.2 Cognitive Models

These models have been devised by psychologists to explain human mental activity by an analogy with computer processing. It is important to remember that models are only an abstraction; the complete story of how the human machine works is much more complex. Cognitive models, however, are useful because they illustrate the advantages and limitations of people, which we need to take on board during RE. Human cognition will be explored using an information processing model based on research by Card, Moran and Newell (1983). All processes require input from the outside world. This concerns *perception*, the process of receiving information from the outside world, while *cognition* is the mental activity we describe in everyday terms as reasoning, problem solving, thinking and learning. The boundary between the two is blurred because as we receive information, we also interpret it.

According to the model each perceptual sense has a processor and associated short-term memory (STM). These memories form the input and output buffers of the human system, storing abstract images in visual short-term memory and sounds in auditory short-term memory. Each memory is associated with a sensory processor. The sensory processors analyze the contents of their memories and pass the resulting information to the cognitive processor for identification of the sensory input. The overall schema of the model human information processor is illustrated in Figure 2.1.

The capacity of sensory short-term memory is not clear, but for vision it must be at least the contents of one visual field, i.e. what we can see at any one point in time. The visual input buffer is constantly overwritten because the quantity of data in an image is vast and images change continually; consequently storing even a few images would take a vast amount of memory. This has implications for extracting information from moving images (video or film) as we remember only the high-level outline or gist of what happens but rarely any detail within individual scenes. Furthermore, we rarely report detail of visible actions or objects, so the requirements analyst has to fill in the gaps by careful observation. Similarly, auditory STM has a limited capacity and it is constantly overwritten so we have a similar problem with extracting and remembering detail from speech. To complete the model, the cognitive processor has an associated *working memory* that is used for storing temporary information.

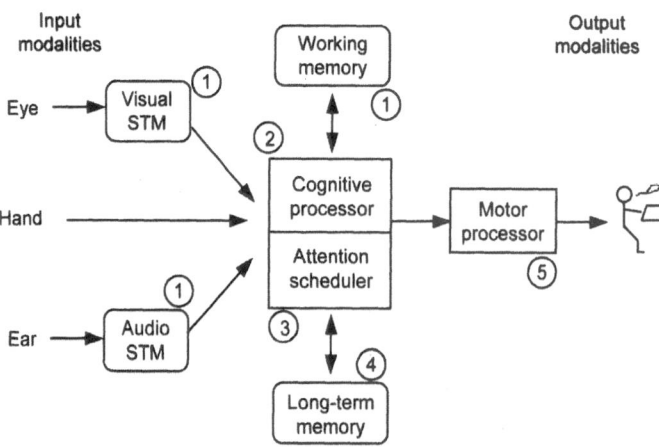

Bottlenecks
1. Capacity overflow: information overload
2. Integration: common message?
3. Contention: conflicting channels
4. Comprehension
5. Multi-tasking input/output

Fig. 2.1 Architecture of the model human processor.

Meaning is generated when information in the input short-term memories is passed on to the central cognitive working memory for interpretation. This is effected by matching the incoming information with past experience and then attaching semantic meaning to the image or sound. The cognitive processor performs most of the actions that are considered in everyday language to be thinking. The results of thinking are either placed back in working memory, or may be stored in long-term memory, or may be passed on to the motor processor. The motor processor is responsible for controlling actions by muscle movements to create human behaviour, for example running, talking, pointing, etc. Speech output is a special case that requires a separate output processor and buffer of its own.

2.3 Speech and Language

The processing of speech and language has important implications for RE. Requirements are captured by speech during interviews, discussed and negotiated during meetings, and documented in natural language as well as in specially designed languages of formal methods and diagrammatic notations.

Language recognition from speech starts by discovering the basic sound units of language and matching them to units of written language, which correspond approximately to syllables, suffixes, prefixes, etc. and thereby words. Interpretation, however, is a layered and integrated approach in which the brain makes use of language syntax (the grammar), semantics (the meaning of words and sentences), and pragmatics (knowledge of the context of communication) to decipher communication. Speech is full of incomplete sentences, inaccuracies and mispronunciations; however, we rarely notice when people make mistakes. This is because we automatically supply corrections to what we hear from memory. So if a speaker misses out a word or doesn't finish it, we subconsciously supply the missing component, for instance, "the eel was on the car" would be heard as "the wheel was on the car" by most people. Unfortunately this means that we can hear things that were not said, and hence misconceptions can arise.

2.4 Memory

Memory plays a crucial role in both vision and hearing; consequently the role of perception in the sense of receiving information, and cognition in the sense of understanding and using external information, cannot be meaningfully separated. This leads to investigation of how memory works and how it is used in the processes of understanding and reasoning. Human memory comes in two varieties: working memory and long-term permanent storage.

2.4.1 Working Memory

Working memory is the human equivalent of computer RAM, in other words the working memory of the central processor. In contrast to computers, human working memory is small and loses its contents unless it is refreshed every 200 ms. Items are not stored in computer memory bytes but in "chunks" of information. These can vary from simple characters and numerals to complex abstract concepts and images. The secret of expanding the limited storage in working memory is to abstract qualities from the basic information and store the abstraction instead.

This concept is best understood by example. Telephone codes may be given in an unordered fashion, such as 01612363311. Such large numbers are difficult to assimilate and remember, but break the number up into smaller units and memorization is easier, for example 0161 236 3311. The effect is to suggest a chunking strategy to the reader. Instead of storing 10 separate digits, the number groups can be stored as whole chunks, reducing the storage required from 10 chunks to three. The more order that can be imposed on the raw data, the better the chunking. Working memory limits our ability to process information. It has many consequences for reasoning and problem solving. Overloading working memory can cause problems, both in terms of quantity of information and time span of retention. Designers should therefore limit the quantity of information provided in a short space of time. People can read detail later; information in requirements meetings and interviews should be kept to the key points. Structuring (chunking) information helps memorization. Categorization, grouping, sorting, using formatting and visual aids all help people assimilate more complex information by facilitating the process of abstraction. Large quantities of dense text should be avoided if people are to understand requirements in meetings. Images are helpful but need to be accompanied by text. Diagrams accompanied by captions and speech explanation help to abstract and summarize essential details of arguments.

2.4.2 Long-term Memory

Long-term memory is the main file store of the human system. It has a near infinite capacity as no one has been able to demonstrate an upper limit on what we can remember. Memory failure appears to be a problem of not retrieving what is already inside our memory. Memory retrieval is a two-phase process:

1. *Recognition*: the initial activation of a memory trace by cues.
2. *Recall*: the actual retrieval of the information itself.

As a consequence, you may often have a fact on the tip of your tongue (recognition) but can't quite remember it (recall). In frequently used memory, both recognition and recall are so quick that no difference is noticed. Memorization and recall are helped by recency and frequency; the more often we use a piece of information the easier it is to learn it. Retrieval of facts from memory can be remarkably fast, especially for frequently used items. Retrieval time for less frequently used information varies; it can be quick, but may be slow especially for older people. Memories are found by a process of "spreading activation" as remembering one fact often helps the recall of other related items. It is rather like a large net of interconnected facts that becomes sensitized by use. One consequence is that rehearsal of points summarized from previous interviews helps to activate participants' memory, while summarization of key points at the end of meetings helps to create an effective memory.

Memorization is usually an effortful process. Most learning is by association, in which facts are linked together to provide an access path. The greater the number of separate access paths, or the more often an access path is used, the easier a fact is to remember. Problem solving or reasoning during memorization creates more links and hence helps recall in the future. This increases the "depth of encoding" which means when we understand something more thoroughly we tend to remember it more effectively. As requirements analysis involves learning how a future system should work, more active engagement by users and designers in problem solving leads to better understanding and memory. More accurate memory means less has to be recalled by looking at notes and documents, fewer errors will result, and understanding will be shared more efficiently.

External Memory

Because we have difficulty with processing information when we are doing several things at the same time, and because memorization is a slow and effortful process, we rely on a variety of memory aids, such as written documents and diagrams. These help our memory because we can scan the document to refresh for the necessary facts. External representations help memory because the act of recording facts as lists and diagrams means we have thought about them and put some structure on the requirements and their relationships. Structuring facts improves memorization and cues effective recall by providing categories and associations. Requirements traceability is the act of imposing a natural external memory structure on a document database so that requirements can be retrieved easily. Forwards traceability provides links to the source of the problem, i.e. the requirements author, source and content; while backwards traceability points from requirements to design solutions (Gotel and Finkelstein, 1994).

Furthermore, notes, annotations and drawings can record new facts arising during meetings, thereby saving us the burden of memorizing them. Not surprisingly, notes and documentation play an important role in RE, and external representations need to be carefully designed to help the RE process. External representations are necessary for providing input to the reasoning process as well as recording the results. Diagrams and structured text notations record abstractions that form models and specifications of the required system, while detail of the domain can be represented as scenarios. If we base external representations on the structure of internal memories then they should fit naturally and improve not only memorization but also recall of information.

Organization of Memory

There are several forms of long-term memory which resemble software engineering notations. This is a good illustration of how external representations have been designed, or may have naturally evolved, to fit with human abilities.

- *Semantic networks.* The basic organization of long-term memory consists of linguistically based concepts linked together in a highly developed network. The organization of human memory is far from clear, although most evidence favours the view that all storage is finally of the semantic associative kind, with several different organizations. Semantic network representations appear in "mind maps" that give informal associations between facts, concepts and requirements, as shown in Figure 2.2; later in the process semantic networks may become formalized in software engineering as entity relationship diagrams, semantic data models, process dependency diagrams, etc.

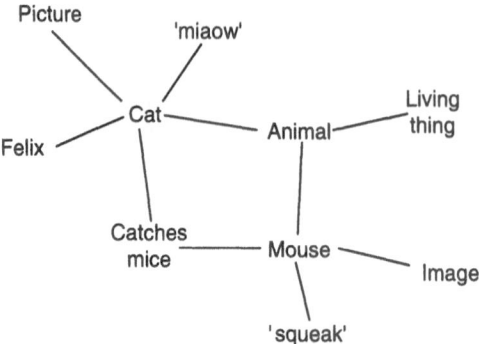

Fig. 2.2 Semantic network with type-instance nodes. This forms a reusable knowledge structure for a common-sense concept of a cat with associated facts.

- *Episodic memory*. This stores events, scenes and contextual information in realistic detail. Episodic memory, in contrast to semantic memory, is composed of more detailed facts about the world, anchored to a particular episode. This memory can store images, sounds and physical detail of an episode that is particularly salient, for instance eyewitness memory of accidents. As studies of eyewitness testimony have shown, however, episodic memory is highly selective and can be inaccurate. Memorization tends to be automatic and linked to the emotions of pleasure, enjoyment or fear. In RE, recall of episodic memory can give rich details of events and scenes for scenarios, but the analyst has to beware that people's episodic memory is highly selective so a sample of scenarios may represent atypical events from a personal viewpoint.
- *Categorial memory*. This is memory of objects and their groupings, familiar as sets, types or classes in object-oriented design. There is evidence that we organize the world not into discrete non-overlapping categories but in a more fuzzy manner, with core and peripheral members (Rosch, 1985). To illustrate the idea, most people have an idealized concept of a bird (see Figure 2.3). A robin fits this core or prototypical image by having the properties round, feathered, sings, lays eggs, etc. In contrast a penguin is a more peripheral member of birds because it does not share all the attributes of the prototype image and it has additional non-standard attributes, such as swims, can't fly. This concept works well for concrete, physical objects, although the situation for more abstract concepts (e.g. goals, functions) is less clear. While people tend to agree about concrete facts taken from the real world, consensus on more abstract categories, such as functional requirements, is harder to achieve. Card sorting techniques (see Chapter 5) help to structure categories, while differences between people can be handled by viewpoints (Sommerville and Kotonya, 1998).
- *Procedural memory*. This is knowledge of actions and how to do things. Computer programs, macros and scripting languages are sequences of instructions or procedures. Procedural memory is held in two different forms: declarative or rule-based knowledge and procedural knowledge (Anderson, 1985). When we start out knowing little about a subject, we acquire fragments of declarative knowledge as rules and facts. This knowledge, however, is not organized, so to carry out a task we reason with declarative knowledge fragments and compose them into a

plan of action. As people become more familiar with a task, fragments of declarative knowledge become compiled into procedures that can then be run automatically. Scripts (Schank, 1982) are a form of procedural memory that encode a sequence of events and their context which we have learned from experience (see Figure 2.4). They represent prototypical "stories" of what we expect to happen in a particular context, for example, when we enter a restaurant the usual sequence of events is to receive the menu, order a meal, eat it, pay for it and leave. Procedures appear in requirements when we ask questions about how something should happen leading to specifications for processes and algorithms. The dichotomy between declarative or procedural knowledge is reflected in the choice of documenting process requirements as rules (declarative representation) or methods (procedures).

● *Analogical memory.* Analogical memory links two sets of domain knowledge that on first sight are unrelated. The concept is best explained by example. Take two domains, astronomy and chemistry, and their knowledge structures, one repre-

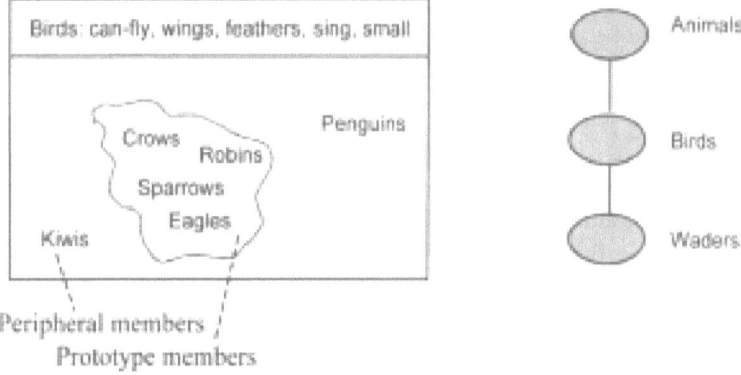

Fig. 2.3 Natural category prototype and peripheral members. Categories are usually organized in three-level hierarchies.

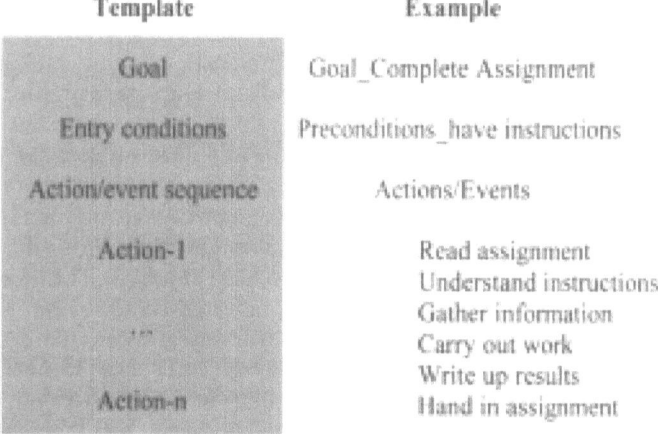

Fig. 2.4 Script version of procedural memory (how to do a task) or event sequences (what happens).

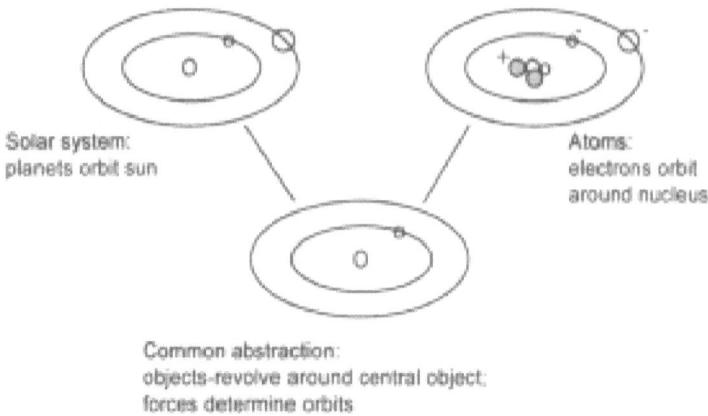

Fig. 2.5 Analogical memory schema and the application of analogies in problem solving.

senting the relationships between the sun, planets, gravity and orbits and the other representing atoms, nuclei, electrons, electromagnetic forces and orbits. The linking analogy is the abstraction of satellites revolving around a central node on orbits determined by forces; see Figure 2.5, after Gentner and Stevens (1983). Analogy is a useful way of exploiting existing memory by transferring it to new situations. Analogical memory is closely related to abstraction or generalization in learning. In requirements specifications we create more general abstract models of more specific examples and scenarios.

Memorization techniques

Memorization can be helped by techniques that add extra semantic cues or associations with the object to be retrieved. Examples are organizing facts in order, using diagrams to associate concepts, keywords to index categories, mnemonics, similes and acronyms. Memorization fails because an access path either decays through lack of use or was poorly constructed in the first place. Similar facts can interfere with recall, so well-recognized access paths that are sufficiently distinct from others are helpful in preventing recall errors. Requirements management tools need to mimic this property by providing several distinct access paths for retrieving requirements.

Memory is one of the critical limiting factors of human information processing. We deal with the complexity of the world by ordering and classifying it. In RE we should support this process by imposing structure on information as soon as possible. We understand and memorize complex information by breaking the complexity down into simpler components using a hierarchical approach. The more structure and categorization we can put into a body of information, the easier it is to learn. Early in requirements analysis domain facts will appear as an unordered jumble. The more structure we can put on categorizing facts and hence the more structured our approach to fact gathering, the easier it will be for us to learn about a new system. This can be summarized as the structuring principle, which is complemented by the principle of consistency:

● *Structure*: facts, information and requirements should be organized in logical groupings and hierarchies that conform with the user's view of the information. Structuring is reflected in standard classifications of requirements (Mazza *et al.*, 1994) as well as in requirements documentation; tools (e.g. DOORS) use classification schemes based on categorial memory.

● *Consistency*: this encourages similar patterns in structures and processes. The more consistent something is, the easier it is to perceive patterns within it and hence to learn its structure and characteristics. Humans are good at pattern recognition and association; anything which helps to establish a pattern will reduce the learning burden.

To summarize, memorization can be helped by enriching information by reasoning during learning. Effectiveness of recall is correlated with the depth of processing, that is, problem-solving effort helps memorization. Structuring information helps memory and creates extra links to retrieve items. Techniques can be used to add extra recall cues, for example keywords, spatial memorization, use of diagrams, combination of visual presentation with text. Consistency of associations creates better contexts for memorization and recall.

2.5 Thinking and Problem Solving

Requirements pose problems of understanding and implementation. Validation poses further problems of understanding how something will work in a particular environment. Problem solving is something we do every day of our lives when we come up against the unexpected. It may be defined as "the combination of existing ideas to form new ideas". An alternative view focuses on the cause: problems arise when there is a discrepancy between a desired state of affairs and the current state of affairs and there is no obvious method to change the state. Simon (1973) laid the foundations of problem-solving theory by distinguishing between the problem and solution space. The former is a model of the problem before its solution (i.e. requirements and domain models) while the latter consists of facts describing the solution (i.e. specifications and designs). Problem solving progresses through several stages:

1. *Preparation or formulation*: the goal state is defined and necessary information for a solution is gathered to create the problem space.
2. *Incubation or searching*: anticipated solutions are developed, tested, and possibly rejected leading to more information gathering and development of alternate hypotheses.
3. *Inspiration*: the correct solution is realized to complete the solution space.
4. *Verification*: the solution is checked out to ensure it meets the goals and is consistent with the information available. In RE this part of problem solving is validation.

The problem-solving process has an important implication for RE: we can only really understand the requirement (a goal state) by thinking, at least partially, about the solution. People use a wide variety of problem-solving strategies, but they are naturally conservative in their approach to problem solving, and adopt the methods they are used to. The more common strategies are reflected in approaches to RE:

● *Problem decomposition.* Complex problems are subdivided into smaller chunks by top-down decomposition and divide-and-conquer approaches. This results in either goal-oriented decomposition to create hierarchies of functions, or sub-systems. Decomposition makes problems easier to solve as each sub-problem reduces complexity and the burden on working memory.

● *Means ends analysis or causal reasoning.* This is reasoning about the cause of an event using a mental model of the domain. Causal reasoning is important for understanding why things happen in a domain and then understanding how a process can be designed to effect change. More formally this abductive, or hypothesis-based, reasoning infers how a set of facts can be explained according to a causal model. It starts with an observation and reasons back to the action or event that caused it.

● *Classification of objects and facts into categories.* This form of reasoning creates categorial memories which underpin object-oriented approaches, for example aggregation of properties to define an object, inheritance of properties in a classification scheme of objects, i.e. lower-level objects automatically assume the characteristics of their higher-level parents. Classification uses inductive reasoning: faced with a menagerie of cows, lions, giraffes and dogs, the observation that some eat plants while others eat meat leads to the conclusion that animals are either herbivores or carnivores. This case is then generalized to other situations where it may apply.

● *Example-based generalization or learning.* This leads to the identification of classes or generalized procedures from specific examples, and can be viewed as the process of abstraction in reasoning. Examples exert a powerful effect on our reasoning because we can identify with detail through our experience. Since scenarios situate examples with existing memory they help in understanding requirements problems. Unfortunately the products of the generalization process, abstractions, are not so easily understood and reasoning with concrete examples as well as abstract models is necessary in requirements elicitation and validation.

● *Deductive reasoning* starts with assertions and discovers new facts by logically examining the relationships or properties that the assertions describe. In RE, we wish to specify the process that will achieve a particular state so we apply a formal method based on logic to help us deduce that the state will be reached. Formal methods use deductive logic to check the properties of requirements specifications to ensure that undesirable states are not allowed to occur; they test for violations of invariants, deadlocks, and that desirable states are achieved (liveness conditions).

Humans, however, do not normally reason formally; instead we use memory and look for patterns and associations. While we reason well in terms of positive association, when negative terms are introduced our reasoning becomes illogical. Take the following problem, which is a classic in psychology (Johnson-Laird and Wason, 1983):

You are given four cards; on each card there is a number on one side and on the other a letter. A rule states that if there is a vowel on one side then there must be an even number on the other. Which two cards should be turned over to prove the rule true or false?

E   K   4   7

Most people go for cards E and 4. Logically this is not correct because the rule states a vowel–even number link and not the converse, so finding a consonant on the reverse of 4 proves nothing. The correct answer is E and 7 because if 7 happens to have a vowel then the rule is wrong. The rule says nothing about consonants (K) and nothing about even numbers always having vowels on the opposite side.

Only too often we tend to look for confirmatory evidence rather than ways of disproving our beliefs. This "confirmation bias" can lead to many errors because so long as the evidence confirms our course of action we do not go looking for evidence to disprove our beliefs. Worse still, we often ignore contradictory evidence when it is available. Confirmation bias can be serious in requirements validation. We will naturally look for evidence that a design will fulfil a requirement but rarely will we try to prove that it will not. This bias means we have to train ourselves to be sceptical and look for negative evidence that demonstrates that a proposed design may not work.

2.5.1 Mental Models

Mental models are important because we construct our view of requirements problems as a set of facts and relationships held in working memory. Unfortunately working memory has a very limited capacity, but this limitation is partially solved by our ability to abstract the essence of a problem and disregard the details. Abstraction is essential for problem solving generally and RE in particular. Our ability to discover the general and ignore irrelevant detail underpins analysis and modelling tasks in RE, but there is a disadvantage to abstraction. Requirements problems often have to deal with "the devil in the detail", so both the abstract and the specific facts are necessary. Abstractions become specifications and designs, while specific facts help to generate abstractions and are used to check designs as test data and scenarios.

The cognitive process of mental models has been explained by Johnson-Laird (1983) and this work has had a wide influence on cognitive psychology. Mental models help explain some important consequences of working memory. To solve problems, several propositions have to be held in working memory. Positive facts are held in memory without difficulty but negative facts pose problems. Representing that something does not exist does not come so naturally and it consumes more chunks. This limitation has several consequences for requirements. First, expression in negatives is always harder and more error prone than positives (e.g. "books should not be allowed out on loan longer than 30 days", compared with "books should be returned within 30 days of the loan date") and double negatives are even worse (e.g. "it is incorrect not to follow the logout procedure"). Secondly, we tend to look for positive evidence to confirm our beliefs – the confirmation bias described above – which means that we look for positive evidence that requirements are correct but rarely look for evidence to prove they are incorrect.

Some of these problems can be alleviated by use of external memory to represent the problem, but there is no substitute for careful thought to ensure that a mental model of the domain is as complete and accurate as possible. A clear mental model for the users of the domain can be promoted by using animated diagrams or interactive simulation to engage the user in active problem solving. This encourages better depth of encoding in memorization. Simulations and interactive models representing external systems are necessary to support users in control tasks in safety critical systems. Such representation helps keep the users' mental model of the external system up to date and promotes situation awareness.

2.5.2 Levels of Reasoning

The way we reason is critically determined by memory. The more we know about a problem the easier it is to solve it. An influential model of problem solving proposed by Rasmussen (1986) has three modes of reasoning according to experience of the domain. If we know little about the problem, previous knowledge and general rules of thumb or heuristics are used. In RE if we have never come across the application domain before then we have to reason from first principles. After some experience, partial problem solutions are stored in memory as rules or declarative knowledge. Reasoning still requires some effort as rules have to be organized in the correct order to solve the problem. Finally, after further experience has been gained, rules become organized in memory as procedures, i.e. runnable programs that automatically solve the problem. Recognition of the correct calling conditions then invokes the automatic procedures (or skills) that consume less effort. People tend to minimize mental effort whenever possible so there is a natural tendency to use skills and to automate procedures with practice. Hence, if we can recognize a previous solution to the problem in hand (e.g. via analogical memory) we will try to reuse it (Maiden and Sutcliffe, 1992). This is the human equivalent of running programmed and pre-compiled knowledge.

Acquisition of skill is influenced by the same factors as memorization. Frequent, regular learning sessions help skill acquisition whereas gaps without practice help forgetting; positive feedback during task performance helps automation, as does presenting a clear model of the task and making the task steps easily recognizable. Redundant feedback only confuses. Skill learning is improved by use of context-dependent learning; this is also important in binding activation of the skilled procedure in the correct circumstances (see also episodic memory).

Skill and automatic processing are important because they enable parallel processing to occur, reduce the need to attend to external stimuli and decrease the load on working memory. The penalty we pay is that automatic procedures are sometimes triggered in the wrong circumstances, even when environmental cues obviously contradict the course of action. Use of automatic behaviour presents a dilemma in matching calling conditions to the correct procedures. In such situations we make errors. We need to consider how users will be reasoning when they are using the computer system, and help the user to construct a clear mental model by building on appropriate parts of the user's experience. For instance, if they know little about the domain (novice users) then the designer should provide ways of learning procedures from rules. The allocation of functions between people and machines, and the degree of automation, should be informed by analysis of how people will reason within a domain. Skilled procedures that are deterministic are candidates for full automation, whereas domains with rule-based reasoning suggest decision support rather than full automation.

2.6 Attention

From the information processing model it should be apparent that there are several input/output channels competing for the resources of the cognitive processor and its working memory. Inputs from the visual and auditory systems compete with other senses that have not been reviewed, such as touch, smell, pain. In addition, the cognitive processor has to find time to access memory and control output to the motor processor and speech buffer. Resource rationing has to occur and, as with a computer, this is controlled by scheduling with interrupts for important events. The control of activity is partly automatic and therefore largely unconscious. This we refer to as "paying attention". If little of interest is happening in the environment we pay little attention to sensory input, as may happen when we are lost deep in thought. The instant something unexpected occurs, for example, a loud noise, our attention is immediately switched to the sensory input. The visual or auditory processors effectively put an interrupt on the cognitive processor.

The demands on our attention are considerable. Take a scenario of a requirements analyst listening to several users while trying to record requirements on a flip chart. Audio input is coming from one or more users at the same time, so the analyst has to separate the speech from different speakers and then understand what everyone is saying. This is done by listening to cues in voice tone and direction, with knowledge of who is speaking and the subject matter. All speech has to be understood with the help of long-term memory and the results held in working memory. Meanwhile the analyst is visually scanning the room to pick up cues about who is attending and who wants to speak next. Visual information also has to be held in working memory. On top of this considerable load, the analyst is also running a cognitive process to interpret what has been said and attempt to find common topics. But the analyst has to run another cognitive process to decide when to intervene in the conversation, another one to plan and rehearse what to say to the users, and then the motor processors have to control the analyst's vocal cords to produce speech. Furthermore, the attention scheduler may also have to control further processes that look at the flip chart, decide what to write and then control the analyst's arm and hand to record a requirement. So up to ten separate processes may be running just to listen to users and say, "Can I summarize the requirement that we must deal with cognition complexity" while writing complexity as a non-functional requirement.

The only way we can deal with so many separate threads is by time-slicing between each process, just as a multitasking computer operating system gives a little processor time to each task. This is called selective attention. Our brain polls each input channel and runs some processes in the background while one is in the foreground. Computers have very fast processor cycles to enable multitasking. People are also good at multitasking but there are limitations. The above meeting scenario is close to the limit. Most people find listening, summarizing and recording very difficult. You have to shut out speech while recording, and even summarizing while listening is not easy. When the demands of selective attention become excessive it is advisable to separate roles between different people, so it is usual to have an analyst who runs the conversation and a scribe who records the summarized requirements.

Selective attention has implications for the representation of requirements that will be dealt with in more depth in Chapter 4. Care has to be exercised that the requirements documents do not produce too many competing demands for attention. For example, attention tends to be diverted by change, hence dynamic media such as film, animation and sound will dominate over static media such as pictures and text. Because speech and video overwrite working memory in a continuous input stream we find it difficult to extract much detail, so important facts need to be recorded in static media: text and diagrams. If too much information is presented at once the attention scheduler cannot cope, leading to information (and working memory) overloading by exceeding our capacity to understand and then either memorize the important facts or make notes.

Another implication is in functional allocation when functional requirements become collaborative tasks between people and computers. The requirements engineer needs to check that the user interface design does not impose impossible demands on human attention by multitasking – for example, asking users to control an operation while monitoring another process. The human ability to ignore the steady state in the environment can lead to poor performance in monitoring tasks. If we have to concentrate on input with little variation, there is a natural tendency to ignore changes and for attention to wander as the cognitive processor polls other channels. Even worse, in long monitoring tasks fatigue may set in, with the result that we miss significant events in the environment.

2.7 Motivation and Arousal

Motivation and arousal are important psychological variables because they influence our choice. Requirements engineers need to specify products that not only do what the users want but also motivate them, either to purchase the product itself or to make the appropriate choice and buy the product in e-commerce applications. Motivation is closely related to user goals, which appear in many RE methods.

2.7.1 Motivation

Motivation – the internal will of an individual to do something – can be influenced by physiological factors (e.g. hunger), psychological factors (fear, sleepiness) and sociological matters such as companionship and responsibility. Motivation affects task performance, decision making and attention. It can be decomposed into *arousal*, which tunes our senses to attend to certain stimuli; *intrinsic* motivation, which reflects the individual's own will; and *extrinsic* motivation, which is linked to properties of a particular resource. Of course these variables interact; for instance, if we are hungry (intrinsic) and smell cooking with garlic (extrinsic stimuli) and have had previous good experience of garlic-tasting food (arousal effect by priming) then our motivation to seek out the food will be increased.

Intrinsic motivations can be ranked in a series of satisfiability (Maslow *et al.*, 1987), from basic survival needs (food, shelter), to reproductive drives (sex, find a partner), curiosity (learning), individual self-esteem (job satisfaction), societal influence (power, politics, possessions) and altruism (benefit for others). Once basic motivations (e.g. food) have been satisfied, motivations for self-esteem, curiosity and power come into play in subtle combinations that marketing specialists are forever trying to guess.

Motivation is important in task design when the designers should try to motivate users by giving them the appropriate level of interest, responsibility and reward for performance. Motivation is a study in its own right that cannot be dealt with here; for further study the reader is referred to Maslow *et al.* (1987).

2.7.2 Arousal

Arousal almost equates with excitement. Arousal is important in design because it is influenced by the content of applications, especially multimedia, and by aesthetic aspects of user interface design (Sutcliffe, 2001). Arousal is poorly understood. Arousal is bound up with our good/bad reaction to content, termed "valence" by psychologists. Dangerous, threatening (chased by a tiger), gory (mutilated body) and erotic content all increase arousal, more so than pleasant images (e.g. flowers, sunset) (Reeves and Nass, 1996). Arousal also affects memory. We remember events after unpleasant incidents more effectively than events beforehand (proactive inhibition). Moderate arousal helps problem-solving performance and increases motivation, so repetitive, boring tasks should be avoided or broken up with more stimulating material.

The relationship between arousal, motivation and attention is illustrated in Figure 2.6. The links and interactions are complex and each cognitive component has some influence on all the others. The nature and strength of these influences is still a subject of active research, so the following description is a brief summary.

Selective attention controls our response to external events. This is partly an automatic process but there is also conscious direction that is under the control of

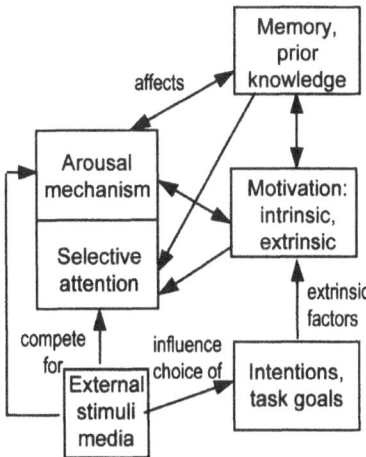

Fig. 2.6 Components of cognitive architecture associated with attention, motivation and arousal.

our will. Arousal interacts with the attentional mechanism by tuning our responses to stimuli as well as making events more memorable (proactive and retroactive inhibition) and increasing motivation to investigate information and take action. Motivation in turn is influenced by our intentions and the task. Goals also influence our attentional mechanism to search for particular information within perceived media input, and memory also influences the process of comprehension. Finally, motivation and prior knowledge affect our arousal and responsiveness to certain content; for instance, memory of a lover increases the probability of our attention to a photograph of him or her.

Control of attention has a complex cognitive mechanism. It is not surprising, therefore, that the connection to design can be mediated by many variables. The design stages that need guidance start with requirements and content selection, followed by selecting appropriate media to represent the content and design of the form of the media, and finally the design of representational detail. These design stages are associated with the second set of variables, starting with choice of material to inform or stimulate. Design of form involves creating media for aesthetic purposes as well as selecting appropriate media for the message, and finally design of attention directing effects.

2.8 Stress and Fatigue

Fatigue may result from continuous mental activity in overlong, mundane tasks and from intense concentration on tasks demanding difficult mental reasoning. In either case rest is required for the human attention system to readjust itself. Fatigue can be caused by repetitive tasks containing no break points. Long continuous tasks should be broken up by rest periods in which the user is allowed to do a mental reset. These break points, called "closure events", should be placed at natural intervals during a task. These intervals could be at the end of an operational sequence, such as entering a transaction record, or a search and replace operation in a word processor. The more complex a task the more demanding and potentially more fatiguing it may be; hence break points should be planned with task complexity in mind.

Task complexity, however, does not always lead to increased fatigue. People find stimulating but demanding tasks interesting. Complexity may hold their attention and delay the onset of fatigue for some considerable time, although highly demanding continuous activity should be avoided because users may be unaware of their tiredness and make mistakes. Mundane, non-stimulating tasks are liable to cause fatigue precisely because they do not stimulate interest and hence fail to hold attention. Such tasks should be automated, but if they have to be manual, a high frequency of break points helps to combat the strain of enforced attention to an uninteresting task.

Fatigue can also be caused by sensory factors. Strong stimuli, such as bright colours, intense light and loud noises all cause sensory overloading because they bombard the perceptual system and demand attention. If exposure to such stimuli continues for a long time the cognitive system will try to ignore the steady state in the environment; however, such strong signals are not easily ignored. This sets up a conflict in the attentional process which can become fatiguing.

2.9 Human Error

Requirements need to deal with human error. This involves detecting errors, containing the consequences of errors and helping users correct the situation. The implications for RE are that functions and tasks should be structured to help users solve problems and that we need to specify requirements that anticipate or prevent users' mistakes. This is particularly important in safety critical systems and this topic is taken up in Chapter 7. The starting point for understanding the psychology of human error is Rasmussen's (1986) model of cognitive processing, and error taxonomies proposed by Reason (1990) and Hollnagel (1993; Hollnagel and Bye, 2000). Human error may occur in three levels of information processing:

- *Skill level*: the problem has been solved previously and we use a pre-formed procedure (a skill) to achieve the desired solution. Cognitive and physical actions are automated and run as programs of pre-compiled skills when the appropriate triggering conditions are recognized. Errors at this level typically concern attention and memory. If we have a poor program stored in memory then it will not be appropriate; alternatively, if attention wanders we may miss out steps in a well-formed program. Functional requirements implemented as algorithms and procedures are the computerized equivalent of human skills. We run these skills by recognizing commands in the user interface, so supporting skills implies user interface requirements to ensure that the right function is invoked for the situation.

- *Rule-based level*: the problem has been encountered before and a partial solution has been memorized as rules, so problem solving is achieved by applying rules to achieve the desired goal. Errors can be caused by memory problems, often misrecognizing the context for applying a rule. The rules may also be inadequate. Avoiding frequency/recency gambling (Reason, 1990) is a key concern at this level, so requirements need to be gathered to enhance situation awareness and decision support.

- *Knowledge-based level*: little or no knowledge exists about how to solve the specific problem in hand, so we have to resort to reasoning from first principles. At the knowledge-based level people suffer from several limitations. First, models of the problem domain are often incomplete, may be biased in scope, and contain inconsistent facts. This tendency leads to many reasoning errors because facts are not available or are incorrect. A further class of errors related more closely to the problem-solving process includes thematic vagabonding (a butterfly mind which does not concentrate on any one topic long enough to solve the problem); and encysting (getting bogged down in detail and not being able to see the wood for the trees). Errors at this level are linked more closely to general abilities of individuals rather than their experience. Requirements are for general decision-support facilities, e.g. note pads, ways of structuring and representing problems.

As we encounter new problems, knowledge-based processing produces solutions first as rules; then, as problems become more familiar, skilled knowledge is created. Errors at the skill level are classified as *slips*, i.e. failures of attention and action in an otherwise well-formed plan, as opposed to *mistakes* when there is some flaw within the planned action (Reason, 1990). At the knowledge- and rule-based levels, errors are more likely to be mistakes, although we should not ignore slips that may have mportant consequences for system dependability. For instance, failure to reset a flag in a program caused a calamitous software failure in the Therac-25 system (Leveson and Turner, 1993); it might have been an attention slip if the programmer intended to set the flag but forgot to do so, or a mistake due to inadequate software engineering knowledge. The taxonomy of reasons for error at each level, which we call "cognitive pathologies", is given in Table 2.1.

At the skill level, errors are caused by failure in attention, when we miss a cue for action, or when we omit an action in an otherwise correct sequence. Rule-based mistakes can occur by the application of "bad" rules or the misapplication of "good" rules; in contrast, knowledge-bound mistakes are rooted in bounded rationality, incomplete or inaccurate knowledge. The right rule is applied in the wrong context or the wrong rule is used in the right context. This gives the "strong but wrong" mistakes in which people carry out action convinced they are right even though the environment clearly contains cues suggesting otherwise. Several nuclear power plant operator failures fall into this category (see Reason, 1990 for case histories). Tasks with high cognitive complexity are prone to mistake-errors. At the knowledge-based level, people suffer from several limitations (Gardiner and Christie, 1987; Johnson-Laird, 1983, 1988). First, models of the problem domain are often incomplete, may be biased in scope, and contain inconsistent facts. This tendency leads to many reasoning errors because facts are not available or are incorrect. A further class of errors is related more closely to the problem-solving process, such as the thematic vagabonding and encysting described above.

We are particularly prone to make errors when time is short and we are under pressure. This leads to frequency and recency gambling (Reason, 1990): when we don't have time to think things through we rely on memory. The most salient memorized procedure for dealing with the situation in hand is retrieved and we run with that procedure without paying too much attention to whether it is the right one. The most salient memories are those which we have used most frequently or have used recently – hence frequency/recency gambling.

Because we get used to normal operation, we tend to pay less attention to what is going on and miss vital cues that failure may be imminent. Worse still, when a failure occurs we have an inaccurate or incomplete mental model of the system, so diagnosis of the problem cause is difficult. This frequently leads to incorrect hypotheses being formed and inappropriate corrective action. The Three Mile Island nuclear reactor accident is a good illustration. The window of opportunity

Table 2.1 Taxonomy of reasons for human error.

Pathology	Countermeasure/Task-support requirement
Thematic vagabonding (butterfly mind)	Task lists, checklists, aide-memoires
Encysting (can't see wood for the trees)	Conceptual maps, issue hierarchies, priority lists, detail filters
Halo effects	Exception reports, history browsers
Confirmation bias	Challenge assumptions, alternative scenarios
Poor mental model of problem space	Problem simulations, design rationales, decision tables/trees, concept maps
Lack of situation awareness: wrong model no model	Training, role variety, simulations, embed regular or update tasks
No/poor validation	Model checkers, testing tools, what-if simulations
Representativeness/availability of data	History lists, hypertext, tracing tools
Belief in small numbers	Data sampling/mining tools, visualizations

that created the pre-conditions for the accident was poor maintenance. A pressure increase in the water heat exchanger caused a pressure release valve to open – a perfectly normal operation – but then it stuck open leading to loss of pressurized water and increase in the reactor temperature. The operators were confused because several alarms went off at once and the system gave them no help in diagnosing the cause. They formed an inaccurate mental model of the system which led them to reduce the supply of cooling water to the reactor, nearly causing a melt down.

This problem of "situation awareness" is difficult to solve. Somehow the human operator has to be kept in the loop of normal operation of the system, so an up-to-date mental model is maintained. Unfortunately the high degree of automation in many modern control systems (such as automated flight control systems) militates against situation awareness. In safety critical systems the requirements engineer needs to gather requirements for decision support and keeping human operators aware of the state of the external system, without inducing either information overload or boredom.

Working memory is another limitation. Only a small number of facts can be held in the "foreground" at any one time, so we tend to form incomplete and partial models of problem domains. We tend to find positive facts easier to hold in working memory than negative facts, so exceptions tend to be overlooked in favour of facts that fit into a normal or positive pattern (Johnson-Laird, 1983). This leads to confirmation biases (looking only for positive evidence), halo effects (only considering the obvious and ignoring exceptions) and poor causal reasoning. The more specialized set of biases, which affect human judgement in numerical estimation problems, has been described by Kahnemann and Tversky (1982) and others. For example:

- *Availability of data*: the frequency of an event is judged by the ease of recall. Recall is affected by frequency and recency of memory use (Reason, 1990), so a biased data set is the norm rather than an exception. Memory of a recent aircraft accident may bias our estimate of how safe flying is.
- *Belief in small numbers*: small samples are judged to faithfully represent the variation in a larger population.
- *Adjustment/anchoring*: once an estimate has been made we tend to be conservative in revising it, a consequence of "stick to the known hypotheses".
- *Representativeness*: our judgement of the probability of an object or event belonging to a particular class depends on how prototypical it is or how well it represents the overall class.

Unfortunately human error is not the exclusive remit of individuals. Although mistakes often originate from individual failings, they may be compounded by social factors. As RE is a collaborative activity and often involves complex socio-technical systems, we need to consider the social dimension of error. At this level errors are more a matter of what we usually term "flawed judgement" than mistakes in reasoning. Quality of design can be influenced by the faith that a group of designers have in their collective product. Estimates of a product's reliability can be biased by social over-confidence. Social pathologies fall into two main classes: group re-reinforcing effects which promote over-confident beliefs; and group value effects where communication tends to bias judgement towards a polarity not observed in individuals. Groups, therefore, have a tendency to be more confident than individuals and to hold more extreme attitudes. A common example is the tendency for group

aggression when individuals underestimate the risk of injury by being over-confident about the group's protection.

It might be assumed that the diversity of opinions within a group will lead to the
consensus decision reflecting the average opinions; however, groups tend to polar-
ize towards more extreme positions and show more extreme adherence to views
than would individuals. This will result in the group taking a more risky stance than
individuals in isolation (Moscovici and Zavalloni, 1969).

A related problem involves group behaviour with relation to dissident views.
There are strong pressures on group members to adhere to perceived common positions, and corresponding sanctions against members who question the group's
stance. This can lead to a "groupthink" mentality in which evidence of flaws in the
group's beliefs are systematically ignored. Linked to this are issues in the wider
social setting, such as the surprisingly strong tendency for people to obey authority
figures (Milgram, 1974) even where the commands are unreasonable.

In summary, it is not safe to assume that workers in groups will tend to find and
correct one another's mistakes if judgement is involved. Groups may not converge
on an "average" position; indeed, the reverse is more likely to be the case. Perhaps
the best way to counteract these effects in safety critical systems is to structure tasks
so that group thinking involves particular attention to safety issues. Finally, many
errors are caused by violations of normal procedures (Reason, 1997) when people
depart from the safe way of operating a system because of poor motivation, lax
management and no incentives to conform, or a culture that does not encourage
safe operation. Many accidents are triggered by poor maintenance as well as hap-
hazard operation; human frailties play a part with poor system design to make the
eventual catastrophe inevitable. The requirements engineer's task is to try and
anticipate those problem causes and plan defences to avoid them. This forms the
subject matter of Chapter 7.

2.10 Social Issues

Requirements analysis in complex multi-user systems needs to be sensitive not only
to cognitive psychology but also to social factors that influence people's behaviour.
Sociology has contributed many insights into requirements engineering from
analyses following the ethnographic tradition of people working with computers
(Goguen, 1993; Luff and Heath, 2000; Sommerville and Kotonya, 1998); however, the
recommendations from ethnography are hard to generalize. Part of the problem is
complexity at the social level. A large number of variables might affect requirements
and user behaviour, such as culture, attitudes, predispositions (or norms), group
identity, etc. The problem is which variables have important bearings on systems
design. I shall single out two key variables that have received considerable attention
in studies of socio-technical systems: power and trust.

2.10.1 Power

Power is determined by access to resources, authority and the influence of control.
Sociological views of power emphasize resource dependencies when one organization needs goods, services or information from another (Emerson, 1962). Psycholo-

gists, in contrast, have modelled power from the viewpoint of personality and dominant-subordinate relationships (Ellyson and Dovidio, 1985; McCrae and John, 1992). Power and trust are inter-related, as power assumes the ability to influence the policy or behaviour of others, which may be exercised benevolently or to the advantage of the more powerful party (Gans *et al.*, 2001). Trust implies the assumption that power will not be used for the disadvantage of the less powerful party. In inter-individual settings, power constrains choice and intent, and influences communication by altering the roles and setting of conversation. For instance, we generally make assumptions of politeness and deference when speaking to people in authority (Brown and Levinson, 1997; Clark, 1996). The more powerful participant has fewer constraints, more opportunities for action, and greater leverage for influencing the decision making of another party. The less powerful party, in contrast, encounters more constraints, has fewer opportunities for action and will have less leverage in influencing others. Power is acquired either by assignment, authority or by experience. Societal roles have power connotations; for instance, judges and policemen have positions of authority and assumed power. Few people question their judgement and most will be influenced by their proposals. Power may also be attributed by membership of organizations, such as learned societies, secret organizations, etc. The experiential aspect of power is a consequence of our perception of actions carried out by other agents and the effects of those actions. The more frequently actions of an agent affect others, and the greater the number of agents affected, the greater will be our perception of an agent's power.

Power influences our judgement by introducing a bias towards making a choice that we assume would meet with a more powerful agent's approval. In conversations, power influences the role we ascribe to other parties and our reaction to discourse acts. An intuitive example is our reaction to commands or perlocutory acts (Austin, 1962). In an asymmetric-power relationship, the less powerful listener will be disposed to obey commands from the speaker, while a more powerful listener is unlikely to obey a command from a subordinate party. In equal-power relationships, commands will be obeyed if they seem to be reasonable; otherwise, they may be questioned. Clearly, other factors will enter into our reactions, such as personality, the force with which the command was expressed, and the contextual setting. Highly skewed power relationships can have pathological influences on communication by preventing questioning and clarification when one party's command or course of action is perceived to be wrong. Other consequences can be biased judgements and ineffective collaboration in tasks, when fear of superiors inhibits a subordinate's performance. However, the merits of power distribution also depend on context. In some domains, power asymmetries are a necessary part of the culture, e.g. military forces or the police. In these circumstances equal power relations could be pathological. Some power asymmetry is a necessary consequence of leadership. A map of influences of power and other social relationships on design is illustrated in Figure 2.7, while power-related pathologies and their design implications are given in Table 2.2.

The implications for power relationships in RE lie in design of the social system and allocation of responsibilities to people. It might be assumed that the computer is always in the subordinate role. However, this may not be true. Computers can possess power by providing access to information, or by their capability to control other electro-mechanical systems. For instance, we all now depend on computers to stop cars with anti-lock breaking systems; in the office we depend on computers to

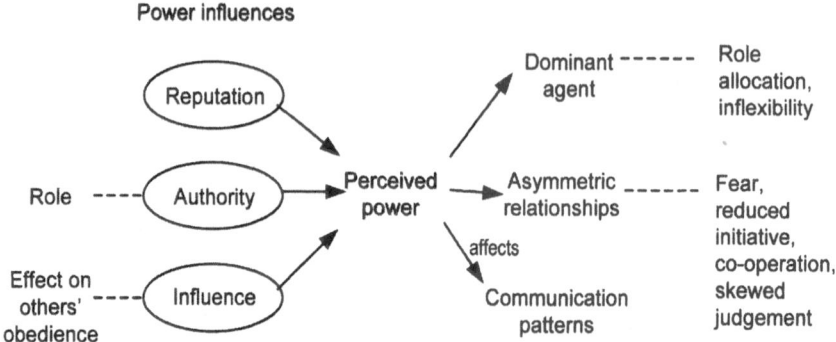

Fig. 2.7 Model of influences on and consequences of power relationships.

Table 2.2 Problems of asymmetric power relationships.

Symptoms/implications	Countermeasures
Commands not questioned even when wrong	Provide anonymous channels to challenge superiors
Decisions biased to appease the powerful	Establish consensus decision procedures, encourage viewpoints
Dangerous events, procedures and systems not challenged	Provide anonymous reporting
Poor collaborative task performance: fear of superior	Reduce fear, improve feedback, praise good performance
Responsibilities abrogated leading to task failure: fear of criticism	Encourage and reward delegation

do most of our jobs of work. Power therefore is strongly associated with the idea of dependency and, as we shall see later, some RE modelling methods emphasize dependency as a means of discovering requirements (Yu, 1994). However, computers can also exercise power in a more direct sense. People ascribe human qualities to computers when speech or human images are portrayed in multimedia (Reeves and Nass, 1996). Indirect power can be exercised via projection of dominant personalities using human images and forceful speech. A more direct manifestation of computer power comes from over-dependence on automated systems for control or providing information; the computer can thus have considerable perceived power in organizations. Such perceptions can lead to pathological effects of stress when computers are designed to monitor, check up on and control human operators.

2.10.2 Trust

Trust is an important mediator of relationships. It has the beneficial effect of reducing the need for governance and legal contracts in managing relationships and is therefore a powerful means of reducing cost overheads in relationship formation and maintenance (Sutcliffe and Lammont, 2001). Doney, Cannon and Mullen (1998) define trust as "a willingness to rely on another party and to take action in circumstances where such action makes one vulnerable to the other party". This definition incorporates the notion of risk as a pre-condition for trust, and includes both the belief and behavioural components of trust. However, different disciplines emphasize different aspects in definitions of trust. For example, economists emphasize costs and benefits, and define trust as the expectation that an exchange partner will not indulge in opportunistic behaviour (Bradach and Eccles, 1989). Psychologists focus on consistent and benevolent behaviour as components of trust (Larzelere and Huston, 1980). Sociologists consider aspects of society, viewing trust as the expectation of regular, honest and co-operative behaviour based upon commonly shared norms and values (Fukuyama, 1995). I have synthesized a common schema from these different views that models trust between individuals, organizations, and also the resources and artefacts upon which they rely.

As with power, trust has an assigned, pre-relationship component: reputation and the experiential component. Trust is assessed initially from information gathered about the other party. Reputation information may come from references supplied by third parties: general publicity, corporate image and brand visibility for organizations, or role assignment for individuals such as professional society membership. A model of trust, developed from Sutcliffe and Lammont (2001), is illustrated in Figure 2.8.

An initial level of trust is derived from reputation. This is then modified by experience, which may reinforce the trust if it is positive or lead to mistrust when negative experience is encountered (Gans et al., 2001). Positive experience results from evidence that our trust in another party has been justified, for instance by the other party's support for one's actions, help in achieving a shared goal, altruistic action, or simple communicative acts like praise and supporting one's case in arguments (Reeves and Nass, 1996). In brief, trust is based on an expectancy of another party that has to be fulfilled; if it is not, then mistrust can quickly set in. Mistrust is not a linear function of bad experience; instead, it can accelerate as cycles of mistrust feeding on suspicion, which leads the other party to question the other's motives in more detail and find criticism where previously there was none. The trust-related problems and countermeasures are summarized in Table 2.3.

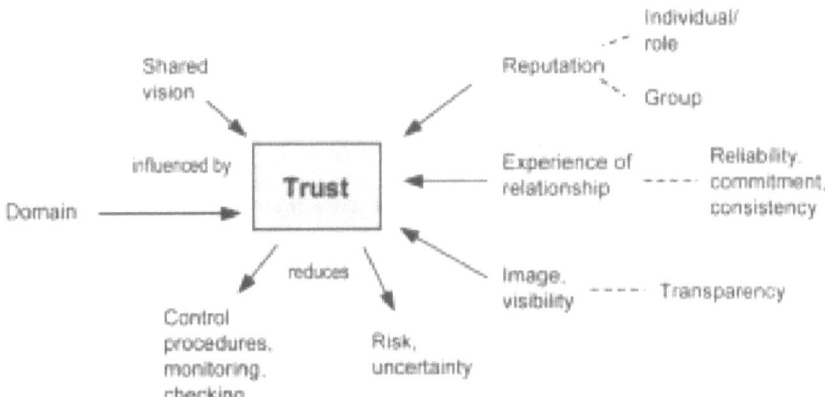

Fig. 2.8 Components of the PLANET (PLANning Enterprise Technology) trust model and their interaction.

Table 2.3 Trust-related problems and countermeasures.

Mistrust problems: symptoms	Countermeasure
Excessive legal governance needed to control relationship	Promote arbitration, clear agreements, third-party regulation of relations
Suspicion about motives of other party: ineffective collaboration	Improve negotiation, clarify shared goals, create agreement for confidence-building measures
Failure to accept proposals and commands from other party	Clarify relationships of proposals to shared goals, explain motivation
Suspicion about information supplied by other party, failure to act on basis of such information	Clarify information, invite third party to validate information
Excessive checking and monitoring other party	Encourage transparency, free information exchange

The need for trust is also domain-dependent. In some systems a moderate level of mistrust can be beneficial to promote mutual awareness and performance monitoring, e.g. rival design teams, internally competitive teams.

The RE implications for trust can be partitioned into trust engendered by illusion, and conscious trust in the whole system. Strangely, when human images and speech are used in multimedia communication, we attribute computers with human qualities and react accordingly (Reeves and Nass, 1996). As a consequence, the variables that influence human trust and perceptions of power can be used to engender trust with the computer – for example, use of praise, polite attitudes, friendly conversational style, smiling, paying attention, supporting suggestions and proposals. Projections of power can be achieved by choice of male authoritarian voice, conversational initiative, image of mature male leaders, and social indicators of power and authority. These effects can be thought of as within-dialogue trust, since as soon as the computer is turned off the attribution is unlikely to persist. Longer range, conscious trust will be built up by reputation and experience, that is, the utility and validity of the computer-provided information (e.g. timeliness, accuracy, comprehensibility), the usefulness of its advice, and its transparency (e.g. explanation facilities, queries, ability to inspect reasons for decisions, etc.).

2.11 Summary

Cognitive psychology can help in understanding important human requirements for the RE process. We interpret the world as we perceive it, by using memory. This is one of the main causes of ambiguity and misconceptions in RE. Two people might both see the same diagram and hear the same explanation from a software engineer but because they have different memories of what the diagram's notation means, each one will put his or her own interpretation on the input. Working memory has a limited capacity, which means we cannot perform many tasks at once. Selective attention allows us to multitask, but the demands of multiple tasks need to be considered when allocating functional requirements to people or computers. Long-term memory has several forms: semantic networks, categorial, episodic, procedural and analogical. External memory uses diagrams and other representations of requirements specifications. When we analyze requirements we engage in problem solving. Our ability to comprehend and solve problems is limited by working memory and attention. Different problem-solving strategies and models of memory can be used to develop methods for requirements analysis and underpin many common software engineering notations.

The model human processor points to some bottlenecks that have important consequences for the way we think and process information in requirements analysis. Reasoning involves forming mental models of the problem. Mental models are a collection of facts and their relationships held in working memory. The limitations of working memory can be partially alleviated by external representations and use of abstraction; however, detail will still be required to validate design solutions. The poor ability to hold negative facts in mental models increases confirmation bias and leads to validation failures. Domain knowledge and reuse of previous solutions can play key roles in solving requirements problems. Human reasoning is not strictly logical; instead it tests propositions and compares facts that make up a mental model. Reasoning is heuristic in situations when little is known about the problem. The early stages of RE when we know little about the problem are the most difficult, yet this is when we have to take critical decisions about scoping. Experience leads to the results of reasoning being stored, first as declarative, rule-based knowledge and then as automatic procedures. We can reuse knowledge of previous solutions to help reasoning in RE. Automatic procedures, or skills, have calling conditions. Mismatch of calling conditions and procedures can cause mistakes. Mismatch between memory of a previous application and the current requirements problem can lead to inappropriate reuse.

We make mistakes when we fail to reason properly or apply faulty memories to problems. Other errors, called slips, occur when our attention wanders. Attention is the process by which we time-slice between many different tasks; however, we sometimes fail to recognize cues and lose the thread of what we are doing. This causes slips in skilled behaviour. The causes of human error are complex. Most failures can be traced back through several layers of influencing factors, from the immediate cognitive failing to the influence of poor computer system design, contributions from the environment such as interruptions and bad weather, and finally to poor management and the culture that allowed the conditions for errors to persist.

Human behaviour is also influenced by social relationships such as culture and group identity. Social relationships are complex and many different variables can contribute to our actions and decisions. Two of the more important are power and trust. Power reflects the authority and discipline aspects of systems, while trust is related to commitment and a more benevolent view of others. Both trust and power have an initial level that is set by information we gather before the relationship starts and an experience component that can change our initial perceptions. The implications for RE are to analyze social relationships and avoid pathologies in complex systems such as excessive use of power or mistrust.

3 RE Tasks and Processes

RE is a flexible process where activities and tasks can be mixed in many different combinations. So while there is no one "cook-book" method for requirements, the basic steps can be described and composed into a process depending on the application context being investigated. However, a single all-embracing RE cook-book method would not be desirable. RE has to fit into a range of different contexts, and integrate with existing development methods such as SSADM, the more recent object-oriented development methods and the unified modelling process. Consequently, this chapter will review the activities in RE as a set of tasks, and then present them as processes for different requirements contexts, i.e. variations according to the application being developed and the starting point of the requirements investigation.

Tasks are a set of actions that achieve a goal. The term will be familiar to readers with an HCI background. Synonyms are activities or processes. This chapter describes RE tasks without going into the detail of specific techniques; instead, the motivation is to provide a "road map" of the RE process and show how this may change according to the application and context in which requirements analysis is being carried out. Although the tasks are arranged in an approximate order of precedence this ordering is not strict. Many RE tasks will be carried out either concurrently or in iterative cycles. One approach to RE is to compose the process dynamically from a library of techniques that can be selected according to the product or presentation being worked on and the goals of the analyst. Hence techniques for creating a class hierarchy depend on the model that exists, if it does, and the viewpoint of the analyst. More detail on context-based process modelling for RE can be found in the work of Rolland and her team (Rolland *et al.*, 1998). This chapter takes a simpler view by investigating different types of RE tasks or activities that have to be carried out on the way.

3.1 Requirements Elicitation

For the most part, the techniques for this task have been borrowed from systems analysis, for example, interviews, observation, questionnaires, text and document analysis (Gause and Weinberg, 1989). Each technique has its advantages and disadvantages:

● *Interviewing*: this is an economic way of gathering information. Interviews may be either *structured*, when the analyst follows a set of questions which are asked in a systematic manner, or *unstructured*, in which case the agenda is open ended. Interviewing can suffer from the analyst's failing to follow up on details; also users may not volunteer their "tacit" knowledge. Interviews are best conducted on the user's home ground in their working environment (Beyer and Holtzblatt, 1998). This has two advantages: first it puts users at their ease, and secondly cues and artefacts on their desk are often useful for illustrating explanations of their work, and demonstrations of their tasks. More detailed advice on questioning dialogues is given in Chapter 4.

● *Focus groups*: this technique is facilitated by group conversation that elicits requirements from a representative set of users. Focus groups are a favoured technique, along with questionnaire surveys in marketing departments, so they do have a political dimension of selling an RE exercise to marketing. The analyst acts as a facilitator, introducing the topic and then soliciting opinions from users. Conversations can improve on interviews as users may build on each other's suggestions; furthermore, disagreements may give early warning of different viewpoints among different stakeholders. Selection of representative users is important in focus groups because a biased selection can give irrelevant requirements; dominant individuals can skew the opinion of the group, so the analyst needs to moderate the conversation to make sure all individuals get the opportunity to voice their concerns. A variation on this approach is to organize expert conversations in which two experts are asked to explain and compare their approach to a problem or domain. Alternatively, an expert can be asked to provide a tutorial to a novice.

● *Observation*: this is less economic in terms of analysis time, but can pay dividends in capturing facts and actions that users do not volunteer in interviews. This technique is necessary to capture physical and spatial information about the layout and context of the system as well as recording physical behaviour. Observation is also useful to get a rich picture of the work environment, how people communicate, informal activities, exceptions and "work-arounds" for normal procedures. Observation can be combined with mini-interviews to clarify points recorded during observation. Unfortunately observation takes a long time and vital events can be missed when the observer is absent. More detailed observation and analysis of work activity is called an ethnographic analysis. This tries to give a detailed view of how people communicate and behave in a workplace, but the analysis is even more time consuming (Luff and Heath, 2000; Sommerville and Kotonya, 1998).

● *Participation*: in this approach the analyst performs some of the users' tasks under their guidance. This technique is a powerful way of learning about the difficulties involved in users' work. Frequently watching an expert perform a task can make it look easy. Trying it yourself soon exposes the complexity. Participation is useful for acquiring knowledge that novices need to learn for a procedure or task; however, it is intrusive on the user's work activity, and it can be difficult to record facts and requirements while the analyst is immersed in doing a task. Nevertheless, participation is a valuable way of eliciting tacit knowledge, hidden assumptions and understanding detail in user tasks.

● *Questionnaires and surveys*: these are useful when users are remote or otherwise difficult to interview. Surveys can gather information on a wide variety of users but the problem is the response rate. Most people are too busy to fill in questionnaires, so response biases can occur. Only people who are less busy and like responding may reply. Questionnaire design is also quite difficult. Questions with answers on 5 or 7 point scales labelled with ranges help to quantify results, for example:

Is the advice on questionnaire design clear? (please circle one number)

| 1 | 2 | 3 | 4 | 5 | 6 | 7 |

very ambiguous   very clear

Some open-ended questions should be added so users can give their own suggestions, comments and criticism. Avoid double topic questions and make sure questions are crystal clear, otherwise the answers can be garbage. Always pilot test a questionnaire; you will be surprised at how people can misinterpret questions that you thought were completely clear. Finally questionnaires are of limited use unless ideas are followed up, so this technique needs to be integrated with others.

● *Documentation analysis*: this technique is useful to understand the existing system, but it depends on the quality of existing manuals and user procedures. If the system is well documented then this is a rich source of knowledge about how the system is structured and how it should work. Documents, however, are often out of date, do not give requirements for the new system, and rarely capture the exceptions and errors in a system. Nevertheless, documentation analysis can give a good picture about how a system was designed.

Generally a mixture of all six fact-gathering techniques is necessary, although interviews remain the most popular technique. Techniques from knowledge acquisition such as repertory grids and protocol analysis have been employed, but there have been no systematic investigations into the merits of different fact capture techniques, apart from a preliminary comparison by Maiden and Rugg (1994). An interesting emergent area is the use of ethnographic and associated "observational" methods (Goguen and Linde, 1993; Luff *et al.*, 1993); however, these have failed so far to deliver explicit guidance for fact capture or analysis, leading software engineers to propose their own "quick and dirty" approaches (Hughes *et al.*, 1995).

3.1.1 Scoping

Requirements frequently start with a vague statement of intent. The first problem is to establish the boundary of investigation and, *inter alia*, the scope of the intended system. Unfortunately, this is rarely an easy process since users often don't know exactly what they want, and knowledge about the intended system is vague. Scoping tends to be an iterative activity as the boundaries become clearer with increasing understanding of the domain shared by all the stakeholders. However, the process is poorly understood. For general scoping, enterprise modelling (Kirikova and Bubenko, 1994) provides a way of describing the business context to discover requirements in the large, i.e. goals, aims and policies.

If few goals or new ideas exist, brainstorming methods, such as KJ or Idea writing can be employed to generate new ideas about what is required. These techniques are carried out in workshops in which participants are asked to think of new requirements and system visions. Brainstorming has three phases:

1. *Generating ideas*: state the terms of reference and then get the participants to create potential solutions for the problem. People do this more efficiently in parallel

so ideas are best written down rather than spoken. Their responses are recorded on small cards and can be swapped among workshop participants who are encouraged to build on the idea or provide an alternative. The act of writing also helps to crystallize what each person wants to say.

2. *Synthesizing ideas and establishing commonalties*: many ideas will be variants on a theme, so groups of similar ideas need to be created. This can be done by writing ideas on post-it notes and then sticking them on to flip charts or wall posters. The groups will need to be seeded with a topic on each poster and then reviewed after people have placed their ideas where they think they belong.

3. *Prioritizing ideas*: ideas need to be ranked to eliminate poor solutions. Judgement criteria need to be agreed, such as ambiguity or clarity of the idea, feasibility for implementation and possible cost, and then the participants are asked to vote for their favourite ideas. The requirements are ranked in priority order and exceptions where the participants don't agree on ranking should be discussed. Bimodal distributions in rankings (i.e. 50% like the idea, 50% don't) may indicate different stakeholder viewpoints that need to be explored.

3.2 Analysis

Requirements analysis follows two general approaches:

1. *Top-down decomposition.* This creates functional models which were preferred in systems analysis methods such as structured systems analysis (De Marco, 1978; McMenanin and Palmer, 1984). The problem with top-down approaches is positioning. Good analysts naturally find appropriate boundaries; poor ones don't.

2. *Bottom-up or event-driven analysis*, which creates a behaviour/structural model favoured in object-oriented methods, such as use cases (Jacobson *et al.*, 1992).

In many cases it is useful to combine these two approaches. Goal analysis is better for analysis requirements from the user's viewpoint and requirements for new systems that do not exist, while event analysis is better for understanding existing systems, and behavioural requirements.

3.2.1 Goal Analysis

Goals may be analyzed by decomposition and modelling in goal trees that express the hierarchy of intentions, while relationships between the goals describe whether the goals contribute to each other or may conflict (Figure 3.1).

Alternatively, goals may be analyzed in functional models, as in data flow diagrams from structured analysis (De Marco, 1978). For problem analysis, soft systems methodology (SSM: Checkland, 1981) gives a means of informal modelling and an analytic approach to discovering problem-oriented requirements. SSM encourages recording problems in the root definition of the system that describes the problem and its context in a succinct paragraph. Furthermore, SSM draws attention to "wicked" problems that have multiple complex causes that are not always apparent in surface symptoms. Rationale-based techniques are also appropriate. These structure the results of analysis in hierarchies of graphs linking goals with

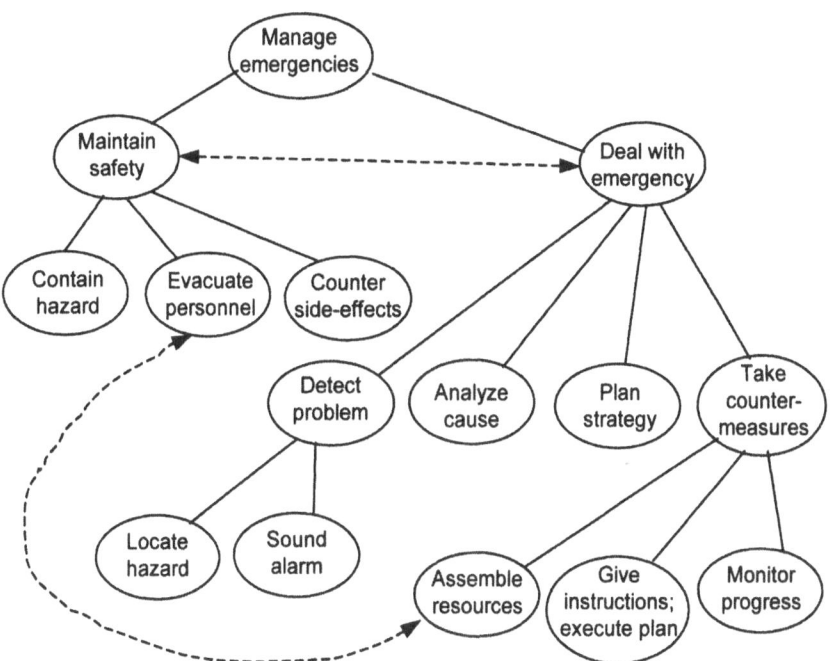

Fig. 3.1 Goal tree, showing relationships for support or conflict between goals as dashed lines.

potential solutions and supporting arguments; see gIBIS (Conklin and Begeman, 1988) and QOC (MacLean *et al.*, 1991).

Requirements may be recorded separately as a list or may be implicit in the models created during analysis; for example, an entity life history records the behavioural requirements of the system. Potts *et al.* (1994, 1995) provide a means of goal-related analysis which uses scenarios to discover obstacles, or potential problems caused by external agents that the system has to deal with. From obstacles, goals for maintaining, avoiding and repairing situations can be elaborated. Other goal decomposition methods follow a taxonomic approach and attempt to analyze goals in the context of domain models (Sutcliffe and Maiden, 1993b).

Goals, aims, objectives and so on, are analysed by recursive "drill down" to explore goals at increasing levels of detail. A goal hierarchy (Loucopoulos and Karakostas, 1995; Bubenko, 1993) and relationships between goals are elicited by top-down decomposition of the problem. Higher-level goals, such as policies, aims and objectives, are all intentions but the analyst cannot say how they will be achieved directly. For instance a high-level aim might be "to increase customer satisfaction". This intention can only be understood in terms of lower-level goals that unpack the aim as "to reduce processing time for complaints", "to improve personal service in ordering" and "to reduce lead times for delivery". As goals are decomposed they approach a level of detail where they can become unpacked into how they may be achieved. This leads to specifications for how goals might be achieved either by procedures, rules or by describing the states that the system should achieve and the constraints on the processes that realize the goal (e.g. Van Lamsweerde, Darimont and Massonet, 1995).

3.2.2 Event Analysis

Another approach to analysis is to focus on events rather than goals or functions. Use cases and object-oriented methods encourage analysis of events and conceptual modelling of the objects and agents. Functional requirements are not immediately explicit; instead they arise as definitions of services that agents (usually users) require.

Events inbound to the system imply a need for a system response or a functional requirement. The event output from the first input process is traced to the next system function by asking the questions "where does the output from the process go to?". This analytic approach builds up event process chains from input through to output and uncovers functional requirements on the way. Processes are identified until eventually output is produced which goes back to the external world. IPO charts (input-process-output) have been used to record event dependencies for many years. The questions in each direction can be summarized as follows:

● Inbound events from the user or the system environment: "what process or function is responsible for responding to this event?" and then, to discover requirements (for validation), "how does the input function know whether the inbound event is acceptable or not?".

● Outbound events from the system to its environment or users: "who or what is the destination of this event and why do they want it?", followed by "which process is responsible for creating the output and what does it need to produce it?".

Event tracing can be carried out in the opposite direction, by identifying agents or users who need system output and inquiring what generates the necessary output.

The outcome of event analysis is to scope the system in terms of its inputs, outputs and major functions. These are recorded in a context diagram which shows the system boundary, major processes and event flows (Figure 3.2).

More detailed event analysis has been researched by Jackson and Zave (1993), who propose techniques for establishing the system boundary by examination of the intended system's obligations for responding to real world events.

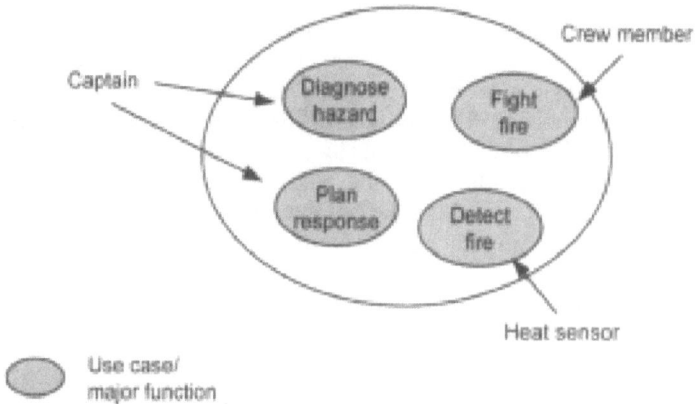

Fig. 3.2 Context diagram for an emergency management system, showing events, major system processes (use cases) and the source/destination of event flows.

3.2.3 Analysis Techniques

There are several analytic techniques which have been applied to systems analysis and RE. These approaches come from the knowledge acquisition literature; and the more important are card sorts, repertory grids and laddering.

- *Card sorts*. This technique is useful for classification and can be applied to a wide variety of problems in RE, such as events, actions, objects, agents, requirements or scenarios. The users are given cards with facts (or requirements) written on them and asked to sort these into categories. They are asked to label and give a rationale for their categories. An expert may also judge the similarity of the categories created by individual users to derive a common categorization. This can be done by discussion and negotiation among the users themselves, although it will take longer to arrive at a consensus.

- *Repertory grids*. This technique is a close relative of card sorting but it systematically analyses membership of categories by comparing the commonalty and differences between attributes of an object (Kelly, 1963). Users are given a set of attributes with which to rate the similarity of the objects. Each pair is judged and rated on a similarity scale then all the objects' ratings are plotted on a repertory grid to show clusters of similar objects. This technique creates a scatter plot-like display and demonstrates whether classes exist among a set of objects and how good the attributes are at distinguishing sub-classes.

- *Laddering*. This is a questioning technique for eliciting knowledge and classifying it in a hierarchy of goals, issues or problems. The analyst asks the user for a high-level fact and probes for the next level fact. The dependency of parent and child facts is recorded as an AND/OR tree. If all the child facts are necessary components for the parent then the relationship is AND, otherwise an OR relationship is recorded. A more sophisticated version of laddering is the "analytic hierarchy" (Ryan and Sutcliffe, 1998) that adds weights to express the contributions that child facts make to their parent.

3.3 Modelling

Analysis and modelling are frequently interleaved to elaborate the requirements as understanding of the problem domain increases. Modelling consumes the output from analysis, structures facts, and represents them in a notation. RE has borrowed techniques for this activity from structured system development methods and conceptual modelling. The semantics of models depends on the purpose or viewpoint of the system being represented. In conceptual models from information systems, three views were common: process/information flows, exemplified by data flow diagrams; data structure, of which entity relationship diagrams are the most common example; and event sequences illustrated by entity life histories. Object-oriented models (e.g. UML) have integrated these views in object class diagrams, with activity sequence and state transition diagrams for detail of procedures.

A more comprehensive RE modelling language is the i* family of models (Mylopoulos et al., 1992; Yu, 1994), which go beyond information system models to represent more phenomena in the world, for example:

- *Tasks*: activities or processes carried out by people or machines.
- *Goals*: goals are divided into those that can be achieved by machine functions or tasks, and soft goals which are non-functional requirements.
- *Resources*: used to achieve tasks by agents.
- *Agents*: people who are responsible for tasks and goals.

Agents, tasks and goals are connected by dependency relationships that express the responsibility, capability and authority for carrying out tasks and achieving goals (see Figure 3.3, after Chung, 1993).

Dependency relationships describe how agents within a system (and their goals) require services from each other in order to achieve their purposes. Dependencies are mapped from the goals that need to be satisfied to activities for carrying them out, and actors responsible for instigating the activities, etc. Rationale-like relationships are

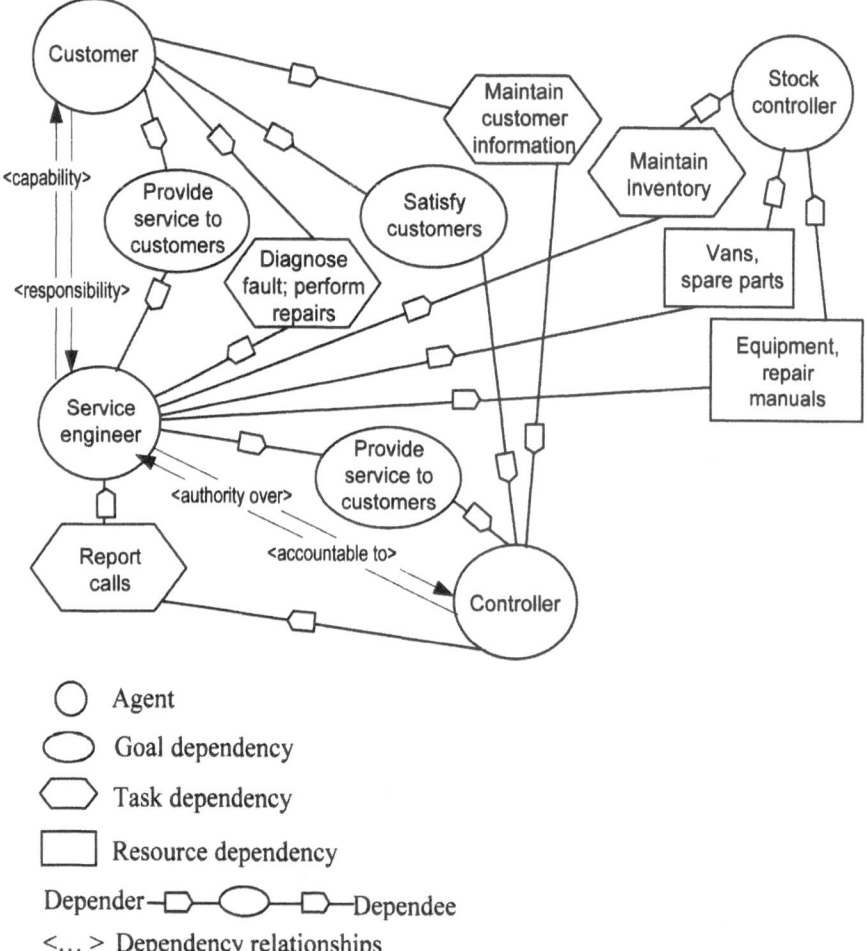

Fig. 3.3 i* model showing dependency relationships between agents, tasks and goals. This describes a map of the relationships between goals, tasks, actors and non-functional requirements (called soft goals), in this case for a service engineer application.

added to show the degree with which a requirement is satisfied by an activity (i.e. a system function). The model can be traced to see who depends on whom (agents) for what (tasks, goals achieved) and with what (resources, objects). Further questions can then be asked about relationships to ascertain whether the dependencies can be achieved. For example, does agent X have the capability to carry out task Y, or does agent X have the authority to carry out task Y and use resource Z? Myopoloulos and his colleagues have modelled i* relationships using first order logic so that some checking can be carried out automatically. In a further variant of i*, argumentation relationships can be added to express how alternative models of tasks and agents can achieve high-level goals. Alternative sub-systems can be modelled with arguments, providing the pros and cons of each sub-system in achieving higher-level goals enabling i* to support strategic decision making for business process engineering and information system planning. Inevitably, models with complex semantics can look fussy and over complex. The trade-off between single models with complex semantics versus many models with different views is covered in Chapter 4. Other enterprise modelling languages in RE have been produced by Loucopoulos and Karakostas (1995) that add business rules with a set of modelling views for objects, actors, and goals; and the ORDIT language (Harker *et al.*, 1993) that represents tasks, actors and their roles.

Models have to be checked to ensure the consistency of the representation. Other checks are tracing relationships and pathways through the model to ensure the system will work correctly. A consistent model should obey the laws of the requirements modelling language that has been used. Modelling languages vary in their rigour and the constraints they impose on how a system can be modelled, so consistency checking can either be a simple task of ensuring that relationships connect the appropriate components, or more complex rules that constrain the types of relationship. Two examples of general consistency checks are as follows:

1. Checks that appropriate components are connected by a specific relationship type.

   model morphology (component (x), relationship (y), component (z);
   schema laws:
   (component type (i) component (x), mandatory);
   (component type (j) component (y), mandatory);
   (relationship type (k) relationship (z), mandatory).<disp\>

   For example, check that all actions are controlled by at least one agent. First a consistency checking tool finds all nodes with a component type = agent, then finds one or more nodes connected to the agent that are actions, and finally whether the relationship that connects the two nodes is of the correct type = control. The tool should be configurable so the input parameters (i) and (j) can be any schema primitives, and the relationship (k) is defined in the schema.

2. Checks whether two components participating in a relationship have the correct properties.

   model (component(x), property(w), relationship (z), component (y), property (v));
   schema laws:
   (component type (i) component (x), mandatory);
   (component type (j) component (y), mandatory);
   (relationship type (k) relationship (z), mandatory);
   (component (x) property (w), mandatory);
   (component (y) property (v), mandatory).

In this example the consistency check is that all agents that are linked to a use case with a <decision> property have an <authority> property. In this case the tool has to search for nodes with a type = agent, and then tests all the relationships connected to the agent node. If any of these relationships is connected to a node type = use case, then the tool reads the property list of the use case and the agent. If the agent does not have an "authority" property and the use case has a "decision" property then the model is inconsistent.

More formal approaches to model checking involve verification using formal methods to determine that constraints are not violated or that the desired states will be achieved. Many formal approaches to modelling have been imported from software engineering (Jackson, 1995; Van Lamsweerde *et al.*, 1995).

3.4 Validation

Validation is a key activity in RE. It involves getting users to understand the implications of a requirements specification and then agree that it accurately reflects their wishes. The current state of the art is walkthrough techniques using semi-formal specifications that are critiqued in a workshop of designers and users. Workshops are composed of the specification author, two or more peer reviewers, a moderator and possibly a facilitator/recorder. The author of the specification walks through the model using a scenario of a typical task, and attempts to show how the system would behave to process the necessary input and output.

The relationships between scenarios, requirements and models in requirements analysis are summarized in Figure 3.4.

Each step of the process is scrutinized by the reviewers to make sure the design exhibits the correct behaviours, and has the necessary resources. Walkthroughs have the merit of early validation of paper-based specifications, whereas prototypes are probably more powerful because users often react more strongly to an actual working system. Unfortunately prototypes incur construction costs and poorly organized prototyping can be detrimental (Attwood *et al.*, 1995). Nevertheless, prototypes in combination with techniques for gathering and evaluating user feedback can be highly effective (Gould, 1987).

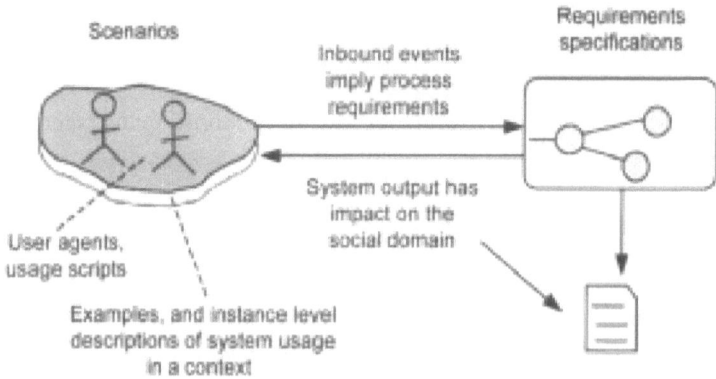

Fig. 3.4 Use of scenarios for validating requirements specifications.

Scenario-based representations and animated simulations help users see the implications of system behaviour and thereby improve validation (Johnson, Feather and Harris, 1992). Simulations help by demonstrating how the system will behave given a set of inputs. When simulations are coupled to a formal representation embedded in a tool then validation can be coupled to verification (Dubois, Dubois and Zeippen, 1997). Verification is checking that the requirements specification behaves correctly and does not violate any of the laws specified by the users. Formal languages are necessary to express the specification in a form to allow verification. This is time consuming, but once it has been done, automatic theorem-proving tools can be used to check that constraints are not violated, the system never reaches an unstable state, it does not get deadlocked or always exhibits desired behaviour given a set of input events (liveness conditions) (Heitmeyer, Jeffords and Labaw, 1996; Dearden, Harrison and Wright, 2000). The boundary between verification and validation becomes blurred when simulations are tested with different combinations of input events. This is the RE equivalent of program testing. Different combinations of input events are run with the simulation and the system behaviour observed, possibly by informal diagrams, to ensure it behaves correctly (verification) and the output makes sense to the user (validation).

The inquiry cycle (Potts *et al.*, 1995) takes a less formal approach to validation by comparing scripts of imagined real world behaviour against the required behaviour in a specification. Obstacles are posited that may derail the system from achieving its goal. For instance, in a library circulation system a professor may ignore a fine notice and refuse to pay fines on overdue loans. In this case an obstacle in the social system means the system's output (a fine notice) does not have the desired effect. Different obstacles are tested against the specification of the system that assumes all fines will be paid, possibly leading to a new requirement to write an official letter to the professor enforcing the fine or putting a block on further loans to that individual.

Overall, the process of validation is poorly understood. Explanation is an important yet often neglected component, and further research is necessary to discover how explanation, representation and users' understanding of system specifications interact. Some research in explaining complex requirements has demonstrated that a combination of visualization, examples and simulation is necessary (Maiden and Sutcliffe, 1994; Carroll *et al.*, 1994a). Scenario-based representations and animated simulations help users see the implications of system behaviour and thereby improve validation (Johnson *et al.*, 1992); furthermore, early prototypes with scenarios are a powerful means of eliciting validation feedback (Sutcliffe, 1995b). A method for scenario-based requirements validation is described in Chapter 6.

3.5 Negotiation

Requirements are often held by different stakeholders who may have conflicting views, hence trade-off analysis is an essential activity for comparing, prioritizing and deciding between different requirements or design options (Kotonya and Sommerville, 1996). Negotiation is necessary in requirements for several reasons:

● to agree which design options are most suitable for the interests of all the stakeholders;

- to select requirements to be prioritized for development;
- to agree a trade-off between conflicting requirements.

Agreeing requirements is often interleaved with validation and modelling tasks; however, negotiation tasks have a structure in their own right with sub-goals for preparation and the conducting of the negotiation, as follows:

- Structure options and choices that need to be resolved. Choices may be alternatives or selecting a number of requirements from a ranked list.
- Establish judgement criteria. A common set of rules, agreed by the participants, is necessary, otherwise judgements about the merits of different options will be inconsistent. Agreeing criteria may require a preliminary negotiation session.
- Explain the options available and the choice that has to be made. This lays out the problem space for the participants. Each stakeholder should be invited to put forward recommendations.
- Invite agreement, ideally by consensus, or by voting if no consensus is apparent. Arriving at agreement is the difficult part. The structure of options and judgement criteria often holds the key to obtaining agreement.
- Diagnose the causes of disagreement. If consensus was not reached then the options and criteria need to be revisited to see if they can be changed or new options generated. Stakeholders who have not signed up to a design will be a source of problems in the future.

Ranked lists or matrix-based techniques using decision tables can facilitate the process, such as House of Quality (Hauser and Clausing, 1988) and the analytic hierarchy process (Ryan and Sutcliffe, 1998) techniques. More complex approaches such as multi-criteria decision making (Fenton, 1995) do not seem to have been considered in RE. The modelling techniques proposed by Chung (1993) and Yu (1994) for mapping relationships and dependencies between goals, tasks, actors and soft goals (alias non-functional requirements or NFRs), contain some guidance for trade-off analysis. A strategic dependency model (Figure 3.3) creates a map of the relationships between goals, tasks, actors, etc., followed by a strategic rationale model that illustrates potential system solutions for the requirements with arguments for and against them. The i* method and support tools facilitate tracing influences between goals and NFRs, as well as giving active guidance about potential clashes between different types of NFR (e.g. security may militate against ease of use). NFRs become criteria that must be satisfied in a designed solution, so functional requirements should be assessed and prioritized against NFRs using trade-off techniques (Mylopoulos et al., 1992).

Negotiation involves human communication and natural language, in particular the problems of establishing common ground or beliefs that are shared by parties in a negotiation. Different users will have varying levels of interest in the new system and this may be reflected in their commitment to its success. A useful technique is to categorize stakeholders into three classes as follows:

- Primary stakeholders who will actually operate the system. These users will be concerned with functional requirements and usability issues, i.e. they will want systems that are easy to use and learn. Primary stakeholders may also have most to lose from a new system, which may automate their jobs and make them redundant.

● Secondary users who will not actually operate the system but will consume its output and depend on its operation for successful completion of their work. These users are typically managers who need information to control, monitor and tune a process, and have a stake in the smooth operation of the system.

● Tertiary stakeholders are senior managers who rarely consume the system output directly but make use of information for planning and strategic control of the business. Tertiary stakeholders are interested in the role the system will play in fulfilling their strategic objectives, such as enhancing competitive advantage, improving customer service, etc.

Each stakeholder group will have potential gains and losses in the new system, so when the initial requirements list has been produced a stakeholder analysis should be carried out to identify who the winners and losers will be. This is effected by constructing a matrix that compares stakeholders' interests by requirements to record potential gains or losses (Table 3.1). More details can be found in the User Skills Task Match method (Macaulay, 1996).

3.5.1 Managing Negotiation

Conflicts, personal arguments and entrenched positions are the enemies of successful negotiation. The first problem is to separate issues from personalities. Disagreement can become entrenched because of negative attitudes held by stakeholders about other individuals, which allows the issues that may have caused the conflict to become forgotten. The source of conflict in issues needs to be separate from people's emotions. Interpersonal attitudes are dangerous territory for requirements analysts to stray into, so meetings have to be managed to keep discussion to the issues whenever possible. A common source of conflict is lack of shared understanding. People frequently talk at cross-purposes because they hold different beliefs and do not take the trouble to test their understanding with others. Getting people to explain their views, and the assumptions behind them, is one cure for misunderstanding in requirements negotiation. Guidelines for handling problem conversations are dealt with in Chapter 4. The following techniques for handling negotiation and conflict resolution are drawn from Gause and Weinberg (1989).

● *Handling personal attack.* Being sensitive to the political and personal attitudes held by participants can help to avoid personal attacks by steering sensitive topics away from them. Once attacks have happened, switch the topic or call a break if necessary. The moderator should try to resolve the conflict outside the meeting.

● *Blocking.* This is when individuals react negatively to any proposal without justification, for example "it won't work", "can't be done", etc. The moderator should

Table 3.1 Trade-off matrix for negotiation and selecting design options to satisfy requirements (italicized figures are totalled).

Option	Precise information			Rapid response			Decision support			Total
Show location of all teams	3	1	*3*	4	7	*28*	4	2	*8*	*39*
Show team closest to hazard	8	1	*8*	5	7	*35*	8	2	*16*	*59*
Broadcast hazard location	5	1	*5*	7	7	*49*	2	2	*4*	*58*

challenge these participants to justify their assertions. When the evidence (or lack of it) for a position is exposed the discussion will generally become more constructive.

● *Intergroup conflicts.* Get each group to present the issue from the other group's viewpoint. Group conflict is frequently about a clash of viewpoints, one group simply not realizing what the issue is for the other. Getting people to appreciate each other's viewpoint is half of the battle, but deeper-seated political conflicts may still endure. These have to be resolved by management of the organization. Deal with only the necessary clashes and defer decisions if possible. A cooling-off period always helps and some conflicts evaporate over time. Typical sources of conflict are trade-offs between non-functional requirements:
   - quality versus delivery time;
   - cost versus development time;
   - functionality implemented/solution value versus development time and cost;
   - security versus access;
   - functional complexity versus usability;
   - functional complexity versus reliability and maintainability.

● *Test assumptions.* People often argue from completely different sets of assumptions; shifting the argument from the current clash to the motivations on both sides often uncovers areas for potential agreement. Record assumptions and trace them to their sources to make sure they really do exist.

● *Relax constraints.* Negotiations often fail because solutions are incompatible with constraints laid down by one or more stakeholders. When pressed it becomes apparent that people can live with a less than perfect world, and relaxing constraints allows a solution to be found; for example, a constraint that the system has to fit within 50Mb of disc space might be circumvented by delivering it on a CD-ROM, while extreme reliability requirements could be relaxed if better help lines and software support were offered.

● Try to find *potential benefits* for all stakeholders: the win–win condition. Boehm *et al.* (1994) suggest several heuristics for structuring successful win–win negotiation for requirements, such as finding benefits in reduced time or better service to users who would suffer in other ways.

● When *chairing* negotiations, avoid taking sides; the moderator/chair role in a meeting is there to facilitate agreement among stakeholders and not impose his/her view. Control the floor space so that all stakeholders have a reasonably equal chance to air their views. This may involve restricting the turns of more vocal advocates. Summarize positions and establish consensus.

Stakeholder analysis methods in co-operative requirements capture (Macaulay, 1993) help to structure the composition of workshops and to provide a framework for considering requirements from different viewpoints. Social science research on meetings describes roles, desiderata for leadership and managing consensus in groups (Bowers, Viller and Rhodden, 1994). Groups have a tendency to over-confidence and take more risky decisions than individuals, so their bias needs to be considered as well as the potential bias of strong personalities imposing their views.

3.6 Functional Allocation

Satisfying a requirement may not necessitate an automated system, as management action for changing resources, procedures or responsibilities may suffice. The decision about how requirements are elaborated into functions to be automated in software or implemented as manual procedures has received little attention in mainstream RE. However, such questions have been studied for many years in safety critical systems under the title of "functional allocation", which explores decisions to allocate requirements to people or machines and how to design the cooperation between them (Wright, Dearden and Fields, 2000). System-level requirements can lead to three possible implementation avenues:

1. Management implications for requirements that are not amenable to automation – instead a decision is required about resources, for example "Hire more staff to improve customer service".
2. Operational implications require changes to procedures carried out by people.
3. Opportunity for automated support – a computer system could be introduced to implement a required function, or an existing system could be improved to meet the need.

If automated support is required, the question becomes what sort of computerized support is required to help users achieve their goals. Stakeholder analysis methods (Macaulay, 1993) can help evaluate the advantages and disadvantages of proposed solutions for different user groups, e.g. primary users who operate the system, secondary users who receive its output, and tertiary users who are responsible for the system but do not use it directly. Unfortunately, few methods exist for forecasting the potential change to organizations as a consequence of introducing computer systems. Organizational and enterprise models (Harker *et al.*, 1993; Kirikova and Bubenko, 1994) can informally map out the problem while the i* modelling framework of Chung (1993) gives some support for tracing allocations of responsibility for achieving goals to tasks and agents. Ultimately though, assessing the impact of IT still relies on human judgement.

Users' goals which imply some automated support require further analysis to define the functional requirements for supporting users' work. Judgement about whether a process should be automated or not should be taken in consultation with users. Fortunately there is a considerable literature to draw on in functional allocation, which has its origins in safety critical systems engineering (see Wright, Dearden and Fields, 2000). Functional allocation heuristics, based on the ideas of Sheridan (2000) are as follows:

- allocate repetitive processing, high volume data processing, monitoring to machine;
- people are good at recognizing patterns, associative reasoning, flexible judgement, general purpose problem solving;
- allocate communication, decision making and non-deterministic tasks to people;
- monitoring states and detecting rare events should be automated where possible;
- deterministic procedures with well-defined algorithms are suitable for automation;
- less deterministic processes, with heuristics and judgement are suitable for people;

- control systems with unpredictable events need people to be in command although the monitoring may be automated.

Functional allocation indicates different user roles according to the human requirements for task support. In some cases the user is passive and the computer system performs most of the processing. Requirements in these tasks will be nearly completely automated, apart from simple input and output dialogues. In other tasks the human and computer have to co-operate to achieve a shared goal. Requirements for human computer co-operation usually need to be refined to specify the information a human needs to take a decision. For example, the computer provides facilities to sort and rank suppliers by different criteria (price, reliability, location) to help the user make the decision. To give some impression of the variety of functional allocation problems, some of the requirements issues that need to be addressed are as follows:

- *Automated systems*, in which the computer controls the process. These are embedded systems present in a large variety of products. However, the user interface doesn't go away. Humans usually have to monitor automated functions even if they do so indirectly; for instance, the cruise control of a motor car still has feedback that should keep the driver aware of the speed, how to turn off the control, etc. The implications for functional allocation for safety critical systems are returned to in Chapter 7.

- *Transaction processing*. These tasks are common in information systems. The user has to supply information for the system to process, and the system produces information for the user on request. Requirements are elaborated for input validation routines to check data entry and data retrieval for system output.

- *Information systems*. Information systems are frequently taken to mean commercial transaction processing systems; more specifically, they are applications in which provision of information forms the major part of the system's raison d'e-tre. Hypertext, hypermedia, multimedia and information kiosk applications are typical examples. Computer system requirements concern the information content, its structure, presentation, with user navigation and access controls. More information on information requirements analysis can be found in Sutcliffe (1997).

- *Decision support*. Requirements for these tasks need to be elaborated to answer questions about what information the user needs to take a decision, how much background information is required, what options should be provided, etc. Decision support requirements need a model of the decision-making process. The requirements may vary from a *passive* system which provides information and lets the user make the decision; to *semi-active* systems which provide the users with facilities to pre-process information by filtering, ranking, and sorting options; and *active* systems in which the user sets the parameters and inputs and the system contains a simulation which is run to produce a recommended decision for the user to accept or reject.

- *Interactive and immersed systems*. These systems cover robot control, telepresence and telesensory systems, and virtual reality. They all share requirements for a complex user interface in which the user is represented in the automated world. The user interface may give a sense of presence in the computerized virtual world, with controls for moving and manipulating objects.

Most systems that are not real time-embedded systems have a user interface. Requirements and functional allocation for the user interface should be driven from a thorough understanding of the user's task. The HCI literature (Sutcliffe, 1995a) has detailed descriptions of task analysis methods that share a top-down goal decomposition approach with systems analysis methods but differ by concentrating more on analysis of human activity. Many task analysis methods are based on goal modelling (Sutcliffe, 1995a), so they can complement enterprise or goal modelling approaches (Loucopoulos and Karakostas, 1995). Task analysis methods help to elaborate requirements for supporting the user's work (see Johnson *et al.*, 1992 or Sutcliffe, 1995a), while an approach integrated with software engineering is given by Lim and Long (1994).

3.7 Processes for Discovering and Refining Requirements

Although most applications will involve the principal RE tasks to some extent, requirements analysis and modelling is influenced by the context of the application and initiating conditions, following four main paths. Each pathway has different implications for scoping the initial set of requirements, and how analysis, modelling and validation are conducted.

3.7.1 Policy-Driven Requirements

Requirements are initiated by senior managers and company executives as policies, aims, objectives and other high-level statements of intent. This source includes visions of the future, such as the famous statement by President Kennedy to "send a man to the moon and safely return him to the earth within this decade". This route, illustrated in Figure 3.5, necessitates considerable scoping activity as requirements start with vaguely expressed intentions and users' wish lists. Policy can be analyzed within the business context by enterprise models. Arguably there are non-RE activities that are pertinent, such as business modelling, value chain analysis (Porter, 1980), competitive advantage theories and business process re-engineering (Davenport, 1993; Hammer and Champy, 1993). Business analysis techniques such as business process analysis, concept maps (Eden, 1988), and critical success factors (Rockart and Short, 1991) are also applicable at this stage; however, proposing a detailed methodology is beyond the remit of RE. The key problem is to model the business to discover opportunities for developing computer systems to enhance competitive advantage. Although some suggestions can be found in value chain models (e.g. Porter, 1980) and case histories of interorganizational system design (Holland, 1995) the link between business modelling and RE is poorly understood.

The methods and approaches for requirements analysis (of businesses and computer systems) in the business community are still largely a matter of intuition. However, some links can be made between business theories and requirements analysis. One example is applying the socio-economic theory of interorganizational relationships, transaction cost theory (Williamson, 1981), to developing requirements for computer supported co-operative work (CSCW) systems. Briefly, Williamson's theory predicts the type of organizational relationship according to the goods that a customer wants to acquire from a supplier. Without going into the complexity of the

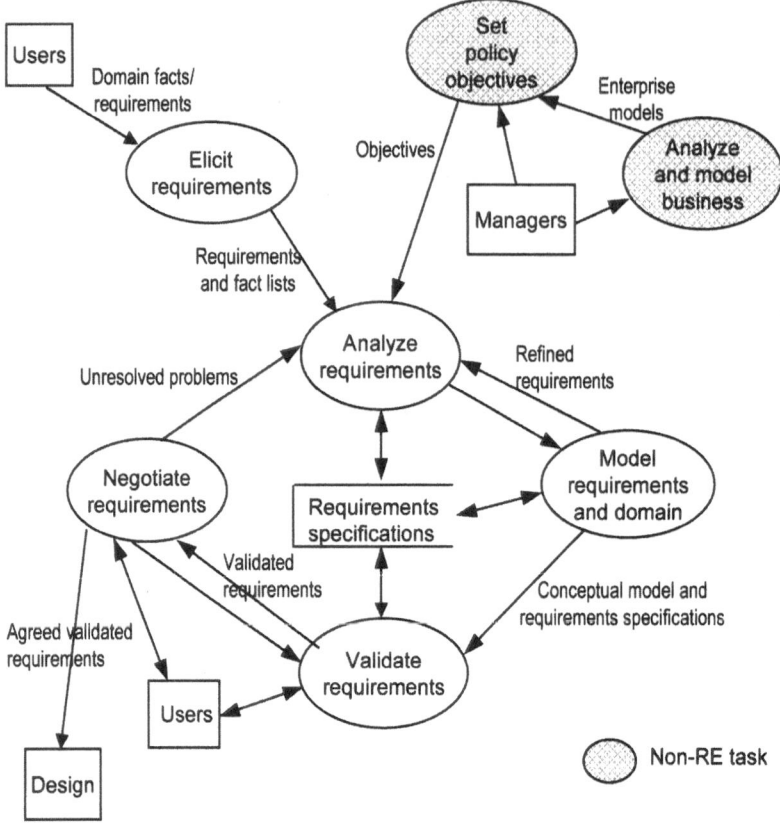

Fig. 3.5 Policy-goal route for requirements discovery.

theory, transactions which involve high unit cost, low volume goods will tend towards a hierarchical, long-term contract type relationship; whereas high volume, low unit cost goods favour market style relationships. From Williamson's theory high-level technology requirements can be predicted to support different relationship types, for example broking and product selection facilities, as found in e-commerce markets; in contrast, hierarchically oriented relationships imply workflow co-ordination, progress tracking, and shared design environments as found in complex aerospace consortia. More details of linking business goals to requirements analysis are given in Sutcliffe and Li (2000), and Sutcliffe (1999a, 2000).

Top-down decomposition is the normal approach whereby policy-level intentions are successively decomposed into goals. Relationships are added progressively as the context of the policy is understood in terms of what has to be done to achieve it (goals) and the implications for people (actors) and their organizations. Modelling goals in the context of how they impact on tasks and the organization is vital not only to elaborate the meaning of informal statements of intent but also to assess the impact of change (Yu, 1994; Chung, 1993). Goals have to be refined as linguistic statements of intent until the stage when the desired state of the system can be described more formally (Dardenne, Van Lamsweerde and Fickas, 1993). Hypertext tools can help to represent informal goal hierarchies (Pohl, 1996), as can conceptual models, for example data flow diagrams, but there is little advice or process guidance for goal-related requirements analysis. Chung (1993) and Yu (1994) provide representations of goals in i* models showing dependencies between goals (both functional and non-functional), actors and tasks, with some guidelines for goal decomposition and modelling. Modelling is an essential precursor for validation in this route as goals cannot be easily understood without contextual detail about how they may be achieved and their relationship to agents and processes (Sutcliffe and Maiden, 1993a; Loucopoulos and Karakostas, 1995). The policy route converges with other RE pathways for negotiation. Once goals have been decomposed to the stage when the desired state of the system can be described, at least informally, the first-cut decisions on use of technology can be made. Some goals become functional requirements, while others have implications for management alone (e.g. decisions about resources, organization, and human activity).

3.7.2 Problem-Initiated Requirements

In this case an existing system contains a problem, and the user's need is to specify a solution. Scoping is necessary but less critical as the observed problem defines the initial scope for investigation. Problem-driven requirements imply that first, the cause of a problem needs to be diagnosed and, secondly, requirements to fix the problem are specified. Two approaches can be taken to problem analysis:

1. *Event-driven diagnosis*, by tracing the events that caused the problem and led to system failure. This approach is derived from the safety critical literature (see below), and is more useful when the problem can be pinned down to an observed failure.
2. *Model-based analysis*, when the problem is harder to identify in terms of a single event.

Event-Driven Problem Diagnosis

In this approach events that led up to the observed failure are traced back in time to tease out the dependencies. Problems frequently have many causes, so it is important to understand all the contributing factors. Two techniques are *event tree analysis*, which follows the pathway of possible events that led up to the observed failure by tracing through a scenario or system task; and *FMEA* (failure mode-event analysis) which constructs a state-based model. Both approaches assess the preconditions that allow errors to occur, which may arise from problems in machine failure, environmental conditions, human error and combinations thereof. More detailed problem analysis in safety critical systems may employ Hazops methods (Swain and Weston, 1988). Cause-effect nets and Bayesian belief nets (BBNs) could also be employed, and a method that analyzes requirements for safety critical systems using BBNs has been proposed (Sutcliffe, 1999a; Galliers, Sutcliffe and Minocha, 1999). Cybernetic models and their derivatives (Beer, 1981) present another view of feedback to uncover potential problems; however, these do not appear to have been applied to RE. Problem analysis for safety critical applications and more detail on these approaches is covered in Chapter 7.

Model-Driven Problem Analysis

The system is modelled to analyze potential problem causes that may arise or may be linked to an observed system failure. Most modelling languages can be used for problem analysis, although few give specific advice on how to discover problems. This responsibility lies with the analyst's expertise and domain knowledge. One method that does mention problem analysis is Checkland's soft systems methodology (Checkland and Scholes, 1990). This provides helps in diagnosing business problems via informal models of processes and influences. The system (and by implication the problem) is summarized in a "root definition" consisting of half a page of text that expresses the system's CATWOE properties (Customers, Actors, Transactions, Weltanschaunung (the organization's desired public image), Organization and Environment). Checkland discusses "wicked" problems that do not immediately suggest causality, and emerge from interaction between several complex variables; however, his method gives little analytic guidance for discovering or curing such problems.

Conceptual models can help to identify different causes that may be attributable to people (actor relationships), performance problems (object property relationships, e.g. volumes or transactions), organizational mismatches, etc. Flow diagram tracing for process/event dependency is useful for problem analysis, and detailed process analysis is often necessary to establish causes. Requirements are realized as technical solutions, or managerial decisions to remedy problems in the social system. Negotiation and trade-off analysis are less important activities as the need to fix the problem drives priorities. Validation, however, is important to check that the proposed solution will actually cure the original problem. The "inquiry cycle" (Potts *et al.*, 1994, 1995) links problem analysis with goal modelling by positing obstacles in scenarios of use which the required system must deal with. Requirements goals are then proposed to address problems, correct errors, and maintain desired states. In general, this pathway has received less attention in RE research than the policy route, probably because problem-initiated requirements are seen as modifications rather than implying the need for a completely new requirements specification.

3.7.3 Requirements-by-Example

These requirements happen when stakeholders hear about or see a demonstration of an existing, innovative application. The user's immediate goal is often to acquire new technology, although the fit of technology with their work goals should be investigated. A variant of this pathway is experts recommending a system enhancement to improve the effectiveness of users' work. Scoping in this pathway is finding the fit between the new technology and existing work practices. Analysis is driven by an existing system that is, in itself, a complex and detailed requirements specification. In this pathway the properties of the designed product are known; furthermore, several alternative product solutions may exist. Modelling may be necessary to establish the fit between users and the new system, although analysis and modelling differ from other RE pathways because a requirements specification can be derived from the existing product documentation. Organizational and task models may be constructed to create scenarios for assessing how the new system may fit into existing working practices, and for deciding how the system or working practices may have to be changed. Trade-off analysis is important as the requirements engineer needs to establish the goodness-of-fit between the requirements and the properties of one or more products that satisfy them.

Product demonstrations, simulations and other early delivery mechanisms help gather information for requirements although these are not truly requirements-by-example, but a manifestation of the required system which has been analyzed by other techniques. The process of analyzing requirements from an existing complete product has received little attention. In practice this approach is often followed when purchasing commercial off-the-shelf (COTS) software and application packages.

Product procurement and selection of COTS software are variants of RE by example. The process is shown in Figure 3.6. Procurement-based requirements imply analysis of the usage context to assess the advantages of the new product need before a decision to purchase is taken. Selecting COTS software involves matching requirements against the properties of products that are available, using decision tables and matrices. A filtering approach is advisable to first establish the quality of suppliers and then establish the goodness-of-fit between their products and requirements. Key questions are:

● What is the maintainability of the product? Is source code available or is the client dependent on the supplier for modifications and bug fixes? Vendors rarely release source code.

● How financially stable is the vendor? If there are any doubts, a copy of the source code should be deposited with a secure third party (e.g. a bank) so modifications could be made by others if the vendor goes bankrupt.

● What constraints are imposed by the client's hardware and operating environment? As Microsoft is a dominating force in the industry most vendors have been compelled to develop for Windows; however, there is the perpetual version control problem: Windows 97, 2000, 200x, etc.

● Which requirements are absolutely essential? Any product not possessing these can be ruled out.

● Can the product be customized to match the user's requirements, and if so, what is the extent of customizability? This quality enhances the fit between the product and requirements but it comes at a penalty of more work. Establish how difficult the customization process is and who will do the work. Customization may be more trouble than it is worth.

The requirements procurement approach can be summarized in the following steps:

● Gather system requirements and prioritize them into essential, desirable, optional extras.

● Select a range of potential products that may satisfy the requirements. This entails market surveys of sending requests for information to product suppliers. Initial matching will be done by comparing the high-level functional description and sub-systems of the products with high requirements/system goals.

● Filter out products and suppliers who are not going to be acceptable on non-functional requirements criteria. Start with suppliers and eliminate those who might not provide sufficient maintenance or where there are doubts about their ability

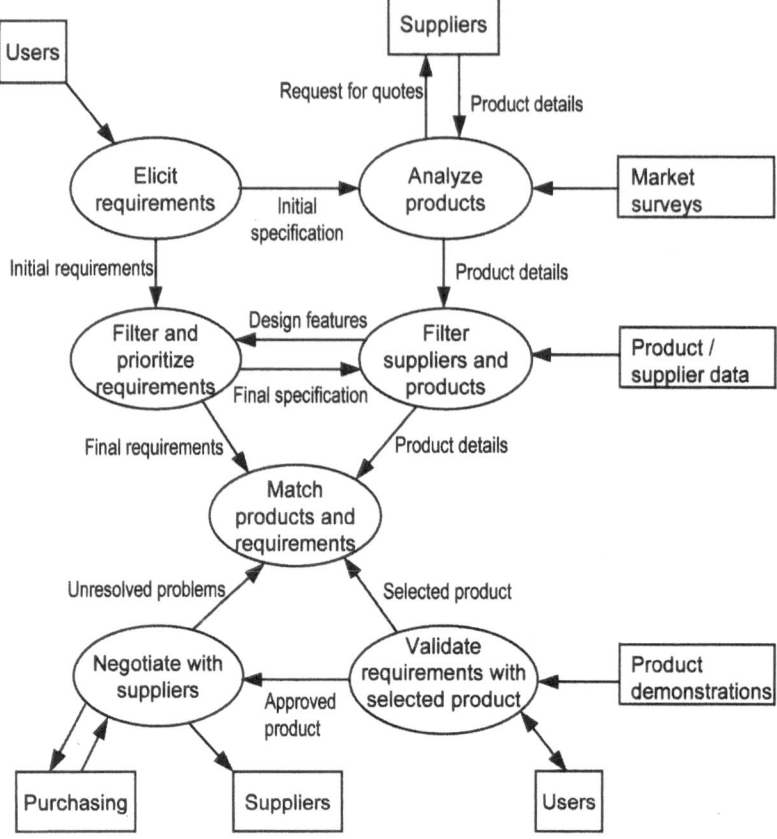

Fig. 3.6 Process model for requirements discovery by example and product procurement.

to deliver (i.e. insolvency problems). Apply further filters to eliminate products that do not comply with the customer's environment: maintainability, interoperability, NFRs. Cost may be another consideration at this stage.

- Attend demonstrations of as many competitor products as possible. Draw up a set of scenarios that cover the tasks the system will have to fulfil. Test-drive all products with a representative set of tasks and users, as vendor demonstrations usually hide usability problems. Product testing is essential to gather data for assessing how well a product will satisfy the user's need, as well as evaluating usability, assessing performance, robustness, reliability, etc.

- List the product functionalities and key non-functional requirements. Cast product functionalities and system requirements in a matrix and assign goodness-of-fit values between requirements for all products being considered (see Table 3.2).

- Sum goodness-of-fit values by product to determine optimal choice. The decision can be made more sensitive to user judgements by adding weightings on each column to express the relative importance of each system requirement.

Some tools have been developed to help COTS procurement, based on matrix-based decision support from the quality function deployment literature (Jacobs and

Table 3.2 Weighted decision table for selecting COTS (or other) products against a set of system requirements. Totals are calculated by adding the weighted figures ($\times 4, \times 2$).

	Selection criteria						
	Functionality		Support		Price		
Option		$\times 4$		$\times 2$		$\times 4$	Total
JAWS	8	32	5	10	5	20	62
HAL	7	28	5	10	5	20	58
Lookout	6	24	3	6	8	32	62

Kethers, 1994). These provide a set of metrics that allow the results of one matrix to
be propagated to another so that a top-down selection analysis can be made. More
sophisticated techniques are multi-criteria decision making which add ranking
decision criteria, range and threshold cut-offs on goodness-of-fit judgements and
many other mathematical treatments to express decision-making criteria (Fenton,
1995).

Matrix comparison techniques can establish trade-off judgements; alternatively,
conformance checks against a product's features can be used. More research is necessary to provide more systematic guidance, and this need will become pressing
with the growth in COTS software. Some guidance on the procurement process is
given in the PORE method (Ncube and Maiden, 1999) but these methods do not
help with the critical problem of judging goodness-of-fit.

3.7.4 Requirements Imposed by the External Environment

Initially these may be high-level statements similar to policies (for example, standards, legislation and regulations to improve safety, reduce pollution) or more
detailed requirements from an external source (for example, to provide audit data
for the chief accountant). The pathway frequently involves non-functional requirements such as safety, security, accessibility (for disabled users), usability (workload
regulations) and so on. Scoping and impact analysis have to establish whether existing systems should be changed, or whether the necessary constraints demand a new
system to conform to the external requirement. Analysis refines the impact assessment once the sub-systems to be changed have been identified. Walkthroughs with
models of the actors, activities and objects can facilitate identification of people,
organizational units and tasks that may be affected. Validation is necessary to
ensure that the system components to be changed will actually conform with the
external requirements. Since these requirements are often non-functional in nature,
trade-off analysis is important for determining how well-designed functions can
satisfy the requirement. Some research has been carried out on modelling NFRs in
the context of functional goals, tasks and agents, enabling dependencies and
options to be inspected (Chung, 1993). Design rationale is also pertinent for modelling the relationship of non-functional criteria to design options (MacLean *et al.*,
1991). Overall, imposed requirements have not received much attention and most
research has concentrated on describing taxonomies of NFRs (Keller, Kahn and
Parna, 1990). At a detailed level, the environment imposes requirements on systems
as a consequence of physical laws of nature. An approach to analyzing and repre-

senting such requirements as formal obligations on the system has been investigated by Jackson (1995). Analyzing requirements that are a consequence of legislative change is an error-prone process as demonstrated by the failure of many government systems, such as social security, tax and passport processing.

Previous automated versions may have to be taken into account when developing a new application. Legacy systems influence RE in a similar manner to requirements-by-example by imposing a set of requirements, or more effectively constraints, on the new development. Requirements for the new system may be constrained by the need to ensure backward compatibility; for instance, with the user interface look and feel for consistency, or with data formats for database compatibility. When a technical system already exists, requirements have to be understood in the context of existing systems as well as from the perspective of stakeholders' goals for new systems. Reverse engineering methods may help recovery of requirements from designed systems, but experience has shown that it is difficult to extract design intent from code, even when some documentation exists (Layzell *et al.*, 1995). Analyzing requirements in tandem with assessing the impact of legacy systems appears to be an untouched area of research.

3.8 RE for Different Target Products

Just as the starting point for RE influences the process model, so does the intended product. Requirements research has tended to assume that the target application will be a bespoke system. This is becoming less common as legacy systems, system evolution and reuse increase their influence. Software is being increasingly developed for configuration, adaptation and reuse as the "middleware" market develops. Requirements for such software are very different from standard, bespoke applications. The different perspectives of requirements for COTS products versus bespoke developments were explored in an RE '95 workshop (Potts, 1995). Grudin (1991) has also drawn attention to the implications of different product types for requirements analysis, in particular to the problems of accessing users when designing shrink-wrap products.

Moreover, different domains have specific impacts on RE, e.g. safety critical applications pose problems not encountered in business applications. Surveying application areas is not feasible given the diversity of computer systems, although the Domain Theory (Sutcliffe, 2002) does take up this challenge and posits a taxonomy of general models for application classes with associated requirements issues. A framework that describes the product's relationship to its intended market for analyzing RE issues is summarized in Figure 3.7. Three dimensions characterize the market orientation from narrow targeted applications to wider ranging products, the degree of embeddedness from user services to automated systems, and from specific products to configurable and reusable components. Different product conceptions imply different requirements analysis activities. For instance, more horizontal market-oriented products will require market surveys to establish requirements; while reuse libraries need a domain analysis. The product framework provides a starting point for investigating such impacts on the RE process.

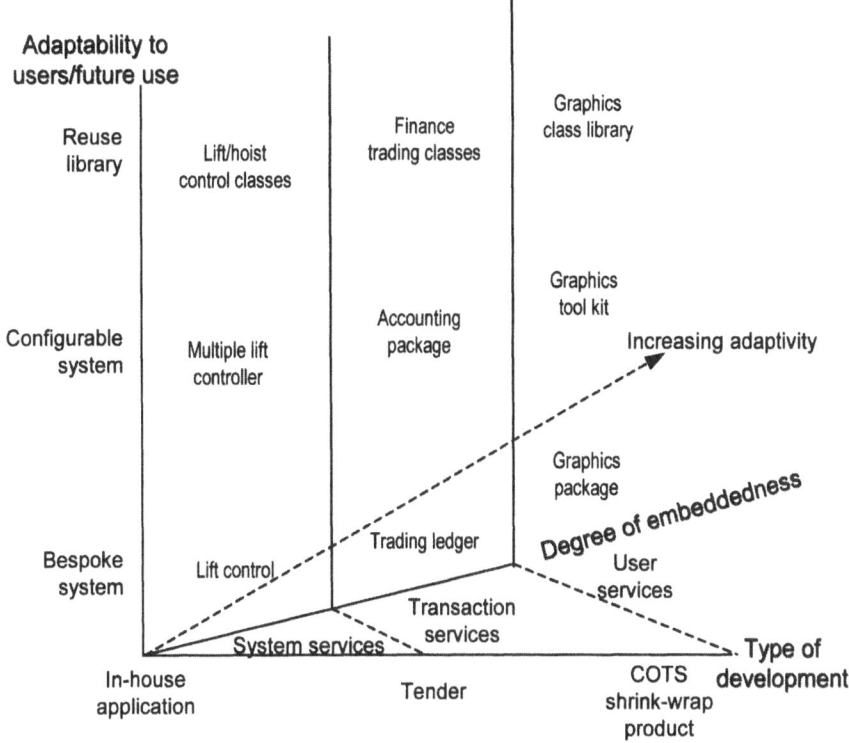

Fig. 3.7 Dimensions of product conceptions in RE.

3.8.1 The Market Dimension

At one end of this dimension, requirements are analyzed *de novo* for a single user and few market considerations constrain the eventual implementation. At the other end, market-led requirements dominate and users are many and harder to identify. Within this dimension there are permutations according to the size and degree of knowledge about the users held by the requirements engineers.

In-House, Bespoke Applications

These imply requirements analysis following the policy route, although the need may also arise from problems in an existing system. Knowledge of the users and the domain is usually available to the requirements engineers. Little competition exists and the user places the contract for the system development with an in-house supplier. The required system tends to be a vertical market type product, and it has to be tailored to fit into the organization's business practices. Requirements for major transaction processing or complex technical systems may be detailed and voluminous, although many in-house developments are for smaller information systems. These are frequently developed by users themselves with the growth in fourth generation languages, application genera-tors and rapid prototyping tools. It is important to involve users in the design process for all systems, but for in-house products there is no excuse for ignoring users because they are in the same organization. Ideally one or more user representatives should be members of the design team; however, putting this recommendation into practice may be difficult. Users do not have the expertise of designers, so they may feel excluded by technical language. User representatives who do learn about the technology can become isolated from their peers in the user department. In spite of these problems user involvement either on the design team or by regular review meetings does improve the quality of requirements capture and design quality (Gould, 1987; Muller and Czerwinski, 1999). Participatory design and stakeholder analysis methods (DSDM, 1995; Macaulay, 1993) are contributions to bespoke driven developments, although evidence for their widespread application is hard to find.

Contract Applications

These are typically large, expensive systems developed for a specific customer. The principal difference with in-house applications is that the starting point is a "request for tender" published by the customer. Requirements analysis may be conducted both by the customer before the tender is published and afterwards by the software contractor who will deliver the system. The implications of different contract types, for example cost-plus or fixed price, may have interesting consequences for RE. Fixed price contracts limit the freedom of requirements engineers to respond to requirements discovered later in the process; furthermore, such implications are rarely considered in the early phases of development. In developments that involve a tendering process, RE may become a collaborative activity between customer and supplier. This helps to make knowledge of the user and domain available to the requirements engineer and common design teams encourage user participation. An example of this approach was Boeing's development of the 777 airliners. Boeing invited representatives of the airlines who had contracted to buy the 777 on to the design team. In the UK defence industry this concept is being tried in common development teams composed of suppliers, customers (the Ministry of Defence) and users (the armed services) (Smart Procurement Implementation Team, 2000).

Products that are open to tender imply that negotiation, trade-off analysis and prioritization are more critical than for in-house RE (see Figure 3.7). Indeed, requirements will unfold and be negotiated as the tender documents are drawn up; moreover, understanding, prioritization and requirements elaboration will continue once the contract has been awarded. Explanation and communication support, such as simulation by concept demonstrators, help the tendering process. Validation is a key concern because of the mass of technical detail, although few solutions appear to have been found (Boehm *et al.*, 1994). Requirements are often expressed in terms of "necessary and sufficient" criteria to which the system should conform. These are similar to non-functional requirements by specifying benchmark criteria for system performance, but differ in that the functional system aspects are described by operational scenarios. For example, a naval warfare system might set down requirements that "the air defence weapon system shall be capable of engaging and destroying no less than five simultaneously launched air to ship missiles at a range of 100 miles, in a range of sea and weather conditions typical of the North Atlantic". The system requirements are not expressed as a list of functions but as scenarios describing what the system should be capable of in a particular context. Necessary and sufficient requirements have become popular with customers who want to procure complex systems, such as the armed forces, because it places the onus of detailed requirements specification on the developer. The sting in the tail for the customer is how to specify a necessary and sufficient set of scenarios for all the operational contexts the new system may encounter. Furthermore, there are many ambiguities in the above example, such as just what types of missiles are anticipated, and the range of weather and sea conditions the North Atlantic may produce now and in the future.

Management of requirements documentation by traceability tools has been a positive contribution of RE research (Ramesh and Dhar, 1992), and some progress has been made on extracting and summarizing requirements semi-automatically from large document libraries (Goldin *et al.*, 1997), but many problems in elaborating and validating requirements for large-scale, technically complex systems are still unsolved.

Commercial Off-the-Shelf (COTS) Products

For these applications, requirements analysis has two phases: first, the developer of the product has to derive requirements for a marketable product; then a customer has to match their requirements to one or more products. COTS products raise different problems for scoping and analysis activities. Users are hard to find, so RE has to rely on surveys. COTS-type products are often driven by a single designer's vision and requirements result from brainstorming (Lubars *et al.*, 1993). Ideally, requirements should be validated with users but commercial sensitivities, such as the need for secrecy, can prevent this. Furthermore, it is difficult to get good feedback from questionnaires and market exercises. One approach is to use mockups and concept demonstrators, then place these in public places or focus groups to observe user reactions (Gould, 1987).

Requirements are often initiated by the marketing department as an entrepreneur's vision or a request for products in either vertical markets (a specific business area such as manufacturing, banking), or in a horizontal market (a more general product need, such as accounting, word-processing, drawing packages). Few methods exist to bridge the gap between marketing and requirements definition, although reports of pragmatic approaches have demonstrated how marketing objectives can be transformed into functional requirements. For instance, marketing certain products is realized by sorting products on favourable criteria in website search mechanisms (Lohse, 2000). Focus groups and market surveys by questionnaires or interviews gather the high-level requirements, but these techniques do not deliver more detailed functional requirements. Requirements discovery is more problematic for shrink-wrap applications as contact between the requirements engineer and prospective users is difficult. Brainstorming and RAD approaches (DSDM, 1995) are frequently used to generate ideas and scope the product. Market surveys can provide further input for scoping and early validation. For horizontal products (e.g. graphics packages) domain knowledge and users may be hard to find; moreover, the context of use may not be known to the requirements engineer.

Decision matrix-based methods such as the House of Quality can be used to establish the goodness-of-fit. General products have to satisfy a wide range of users in the general public, so scoping and analysis for these products is a challenging and little understood process. Vertical market products have a more identifiable domain and set of target users, although customization of off-the-shelf software is becoming increasingly necessary, so this poses problems of requirements analysis with a wide range of possibly inaccessible users. This problem is pressing for e-commerce and web products. One solution is to put prototypes on the web and invite users to give feedback (Hix and Hartson, 1993), but feedback has to be treated with care as the analyst has no control over the sample of users who test the software. Testing products is important and most companies use representative user panels to carry out product evaluations and then form clubs of trusted clients for beta-testing. However, there are few reports of research or industrial experience on requirements definition for shrink-wrap products; for instance, Microsoft's approach appears to be successful but we have no data or understanding about how and why such commercial success is correlated with RE, if indeed it is (Fischer *et al.*, 1993).

3.8.2 The Specific to Generic Dimension

Most requirements research has focused on a specific application in either in-house or contract type contexts. However, many products aim to be more generic and satisfy a range of users and potential applications. The boundary between configuration and design by reuse is fuzzy, as many different approaches are possible. Configuration is generally taken to be "black-box" reuse in which the code is not changed and modules are plugged together to produce a new application. Configuration may also include setting up parameter files (e.g. user profiles) that are used to customize the software's behaviour, format of output, and user interface dialogue. Reuse, in contrast, assumes a "white-box" approach in which the component's code is inspectable and modifiable so the reuser/designer can customize the software to their requirements, albeit by expending more effort. Reuse driven development may also involve the need to develop application-specific "glue" code to co-ordinate a set of reused components.

Configurable Products

These imply some intention by the requirements engineer/developer to produce products that can be customized to different users' needs. The motivation can be either for user interface adaptation so the product delivers appropriate usability according to an individual user profile, or adaptation of functionality to different user groups. The problem is that the range of users and the scope of the domain is more extensive than for bespoke products. This implies analysis not only of a range of required functionality but also of potential user profiles, and the need for configuration tools so that users can modify the product's functionality.

Enterprise Resource Plan (ERP) products from SAP, Peoplesoft, BAAN, etc. are successful examples of this genre. These provide generic applications for commonly occurring industrial sub-systems such as personnel and payroll, accounts, sales order processing, production planning and logistics. Requirements in ERPs are partly set by the vendor as the product embodies business best-practice procedures. However, the processes, databases and user interface can be customized to an extent by switches and parameter files. For more extensive customization a programming language is provided. The requirements process is similar to COTS in selecting the vendor and range of ERP modules, but then it changes to configuration. ERP vendors supply sets of modules that integrate to provide complete business solutions, so the correct mix of modules has to be selected to fit with the business requirements. The configurable sector models that evolved from the ARIS system (Scheer, 1994) into SAP software may be limited to a particular functional area of business; however, they demonstrate the potential for configurable product families based on a thorough, enterprise-based domain analysis. Configuration and customizing can consume considerable resources and require expert consultants who know the ERP products, so the requirements burden is not avoided but merely shifted on to new shoulders: the consultant. Furthermore the limits of configurability mean that often people have to adapt to the ERP software rather than configuring the software to suit people's working practices. This has caused problems when ERP solutions are implemented into organizations with different working practices and cultures (Krumbholz, 1999). Configurability is hard to measure (Keller *et al.*, 1990) so it is difficult to determine the additional effort that may be involved; however, scenarios provide one means of explaining the resource and process implications in configuring products to meet with user requirements (Sutcliffe, 1995b).

The range and ambition of configurable products depends on the defined market need and the power of the customization technology. Ultimately this category includes generative systems whereby user requirements are captured in a domain language and the application is generated automatically. However, research prototypes following this approach have a poor track record of commercial success (Neighbors, 1994). In a more pragmatic but limited scope, fourth generation languages are a successful generative technology.

Reuse Libraries

Reuse libraries share many RE problems with configurable products. The main difference is that the user is expected to build the application by composing it from a library of components rather than tailoring an existing application. The RE challenge is to develop 20/20 foresight, i.e. specify wide-ranging requirements for future uses that can never be completely known. Users and their domain are rarely accessible to the requirements engineer; moreover, the intended use of software components may be difficult to anticipate.

Design by reuse may take two routes: one is to leave the code untouched and deal with intercomponent incompatibilities by bridging functions; the other is to change the component's code to make it more compatible with the new application requirements.

Although RE for generic and reusable products shares the same set of activities with conventional applications, some pathway specializations are necessary. The main pathway for configurable products is assumed to be policy driven, although other initiations are conceivable if users discover examples of new features that they want. The starting point for RE is a market survey or high-level scoping of the product area. Fact gathering, eliciting users' goals and validating requirements can be difficult because users may not be accessible, and even if they are, their future needs may be hard to articulate. A range of functionalities have to be considered, so requirements acquisition needs surveys as well as more standard techniques such as interviews. Validation in a true sense can only be achieved by actual use of the component library, although testing the completeness of a domain analysis may be possible by expert judgement. Validation and negotiation for non-functional requirements may also have to be carried out using indirect techniques such as questionnaires and market surveys.

Analysis for configurable/reusable software has to consider not only the functionality of the product components but also requirements for supporting software that will help users/designers to customize components for new applications. For reuse library products, the pathway for requirements discovery starts with the problem of defining the marketplace for a reuse library. In this case the reuser/designer's task is more difficult, but the payoff is more flexible reuse of a larger number of components. Components have to be retrieved from a reuse library, their functionality understood and interfaces to other modules have to be assessed. If no architecture exists *a priori*, the reuser/designer also has to undertake a general design to determine the architecture and high-level functionality of components, before the reuse library can be searched for appropriate modules. Once these have been located, design involves defining interface compatibilities and possibly customizing code to create the new application. Alternatively an application framework may exist to provide a template for an application area (e.g. user interfaces, data retrieval, communications networks). In this case the analyst determines requirements for customizable hotspots and selects components to fill them (Fayad and Johnson, 2000).

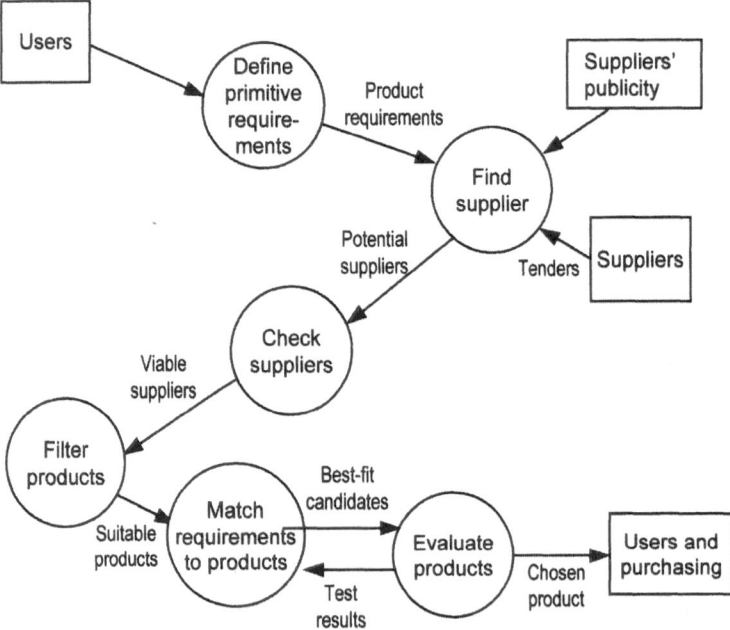

Fig. 3.8 RE activities oriented to off-the-shelf product development and reusable component libraries.

Design by reuse starts with a domain analysis to determine the scope of the new application. The steps thereafter are summarized in Figure 3.8 and lead to the following questions:

● What user groups are present and what are their requirements? What is the range of functionality that should be provided by the new application?

● Which reusable components fit with the user requirements? User requirements are matched to module functions, either by browsing the reuse library for suitable components or by employing a search process.

The next steps progress from analysis to design:

● Select appropriate components from the reuse library; understand their behaviour and interfaces so the design can be planned.

● Design product architecture and tailor the component functionality to fit the results of the requirements analysis.

● Test integration of reused modules and application architecture to ensure that integrity between components and functionality is as desired.

Little research on requirements for reuse libraries has been reported, although some domain analysis methods have been proposed (Prieto-Diaz, 1991; Griss, Favaro and d'Alessandro, 1998) that identify features and generic requirements for an applications sector (e.g. banking) and build reuse libraries for future applications. Domain analysis methods are not very explicit on requirements capture and validation. The general approach is listing functions and possible "use cases" (c.f. Jacobson, 1987) for future applications before proceeding to the design of reusable components. Reusable requirements for a particular domain (jet engine controllers) have been proposed by Lam, McDermid and Vickers (1997). In this approach, requirements were made more generic by creating parameterized rules into which specific values/variables and constraints could be added.

From the requirements viewpoint the new challenges are to discover the range of functionality a library needs to contain, and then to define the scope and granularity of reusable components. For instance, a component in a banking library may perform amortization calculations; however, the compound interest algorithm within it may be reused in a wider range of target applications in finance and elsewhere. Generic requirements models for application classes (Sutcliffe and Maiden, 1998; Sutcliffe, in press) may provide a future solution, although the utility in reusing generic models has yet to be demonstrated in practice. The concerns of abstraction, scope and granularity in requirements for reusable components are complex research topics that are only just starting to be addressed.

3.8.3 The Service Dimension

This dimension characterizes how the target product relates to the real world within which it will be embedded. Interactive systems that support users' tasks, such as decision support, command and control, and ubiquitous applications such as word-processing, imply requirements driven from analysis of users' work, specification of task support requirements and human computer interface design. Task analysis methods, such as knowledge analysis for tasks (Johnson, 1992), can help to model users' activity, but deriving requirements from task models is still largely an intuitive exercise. Some guidance is given for defining task support requirements in integrated methods for HCI and software engineering (Sutcliffe, 1995b; Lim and Long, 1994).

Information systems are driven by business needs to process information, so they may be considered as serving organizations rather than individual users. Requirements for these systems are often expressed as business rules (Loucopoulos and Karakostas, 1995), which are elaborated into conceptual models for information systems. Business applications and small- to medium-scale information systems may be successfully developed using prototyping approaches combined with cutdown versions of structured analysis and design methods (e.g. DSDM, 1995; Crinnion, 1991). However, such approaches are not so effective for complex and embedded systems that interface with other mechanisms and automata, for example software control systems in car engines, aircraft fly-by-wire systems. Embedded systems frequently have safety critical properties (e.g. brake and engine control systems in cars). Since it is important that the behaviour of such systems is rigorously verified against their requirements, more formal approaches to the elicitation of requirements and system obligations are appropriate (Jackson and Zave, 1993). Requirements for embedded systems are frequently safety critical and have exacting standards of accuracy; hence more formal approaches to requirements specification (Van Lamsweerde *et al.*, 1995) are appropriate.

Figure 3.8 summarizes the RE pathways for reusable and configurable products, showing additional activities for surveying and analysis of configuration tools. This involves first deciding the target domain for reuse and then conducting a domain analysis of existing user activities, automated functionalities, etc. A range of functionalities and services are investigated to analyze not only requirements but also all the identifiable components in the domain. Domain analysis methods for this purpose (Prieto-Diaz, 1990) follow systems analysis approaches but are more exhaustive, although there is only rudimentary guidance about what and how much should be analyzed.

Requirements for user guidance and support tools for configuring the target system have to be investigated, with the necessary user/system models, parameter files, and their editing/configuration interfaces. This activity involves modelling users' expectations in terms of how much work they may be expected to do to tailor a configurable system to their needs. Other concerns may be the need for explanation facilities to help users derive the optimal potential from complex, configurable products.

Projections of future usage are hard to achieve since requirements may only be determined within the context of eventual use. For instance, larger design architectures may fit the requirements for a vertical market sector (e.g. financial dealing systems), although components within these architectures may have more widespread target applications – forecasting and amortization algorithms could be exploited in a variety of financial applications and elsewhere. Requirements may also be reused either as generic requirements attached to generalized domain models – loan applications have generic functional requirements for processing loans returns and reservations (Sutcliffe and Maiden, 1998); or as rules with parameters and general variables (Lam *et al.*, 1997).

In reuse libraries requirements also concern non-functional criteria such as ease of construction, portability, interoperability, and maintainability. These in turn become further functional requirements for developing configuration support tools, development environments and software harnesses for integration testing. Reuse in RE has to solve the dilemma of providing material that has sufficient utility to help a requirements engineer without unnecessarily prescribing the developer's view of a new system. Furthermore, reuse can encourage a copycat approach leading to errors, so reuse-led development has to be treated with care (Sutcliffe and Maiden, 1990). Further research is necessary on methods and tools to help developers assimilate and understand the documents and models produced by others. The other key research issue is scaling up to provide significant libraries of generic components for realistic domains.

3.9 Summary

A framework for assessing research and practice in RE is proposed. The framework is used to survey the state-of-the-art research contributions and practice. The framework considers a task activity view of requirements, which commenced with requirements elicitation and progressed through a sequence of analysis, modelling, validation, negotiations and functional allocation. Guidelines and techniques for each task were reviewed. Requirements analysis and modelling are influenced by the context of their initiation conditions, following four main paths. This elaborated different views of RE depending on the starting point of a system's development: the policy route that involves goal modelling and top-down functional decomposition, problem-driven requirements, requirements-by-example, and externally imposed requirements, including legacy system impacts. Another perspective is to analyze RE from different conceptions of products and their properties. Requirements have tended to be considered for bespoke systems; however, market-driven requirements are also common for vertical and horizontal market products. Three dimensions were proposed that influence RE activities. The first dimension described the type of application product, ranging from bespoke to market oriented; the second dimension covered the type of application from real time embedded systems to interactive, distributed systems; while the final dimension addressed the adaptability of the intended product and discussed configurable systems, ERPs and requirements for reuse libraries. Process models for RE within these dimensions were described, and COTS product procurement was reviewed as a specialization of requirements by selection. RE research was examined within this framework and then placed in the context of how it extends current system development methods and systems analysis techniques.

RE has a considerable contribution to make to software development in a broader sense. One of its strengths is an eclectic approach manifest in the range of influences from ethnography, design rationale and scenarios. RE opens the debate about different conceptions of design ranging from the formal, methodical and artefact-based. One potential benefit may be the emergence of a meta-theory of design that acknowledges that applications are diverse and that a battery of techniques is necessary to address different requirements needs. This trend has been noted in specification languages and methods with the argument that appropriate techniques need to be matched to different applications. The pathways and product dimensions proposed in this chapter may render some service in mapping the diversity of issues that need to be addressed.

4 Understanding Requirements Conversations

One of the most important, yet error-prone, aspects of RE is language. Users, analysts and designers have to communicate to discover a mutually agreed solution that the users want and the designers can build. Understanding language is therefore an important part of understanding RE. The following text draws on the science of linguistics and conversation analysis that belongs to the discipline of sociology. One branch of linguistics, pragmatics or discourse, is particularly relevant. Researchers in pragmatics are concerned with how meaning is constructed in context and how conversations are controlled between parties. In RE we need to understand the pitfalls in misunderstandings and develop effective procedures to manage requirements conversations.

4.1 Introduction to Discourse Theory

First we need to consider some background to the process of communication. For that purpose we shall use Clark's (1996) theory that describes the way we establish common ground or shared understanding in conversations. Clark argues that people communicate to establish and achieve shared goals. Language is therefore a prelude to action, and conversations are motivated by a purpose. Meaning develops through a cycle of exchanges between the conversees which establish a *common ground* of shared understanding. Clark proposes a *ladder of action* that moves from initial recognition to comprehension of intent, elaboration of implications, planning actions and finally executing plans. However, meaning is construed not only through language but also via reference to mutually accessible artefacts (e.g. pointing to diagrams or objects in a room), and via shared knowledge of the communication context and the roles of conversees. Space precludes an elaborate treatment of Clark's theory; however, even this précis can be used productively for reasoning about design. As before, the approach is to locate pathologies within a model and to suggest guidelines to remedy such problems. A generalized task model of communication based on Clark's theory is illustrated in Figure 4.1, while the associated pathologies and design guidelines are explained in Table 4.1.

The communication model follows the traditional descriptions of cognitive processes for natural language understanding that commence with the lexical level of phoneme/word recognition, progress to interpretation of the utterance using the rules of grammar (syntax) and knowledge of word meanings (semantics), with final understanding being created by integrating input within contextual knowledge of the conversation setting, participant roles, the conversation history and domain knowledge (pragmatics). In Rasmussen's (1986) framework, communication is gen-

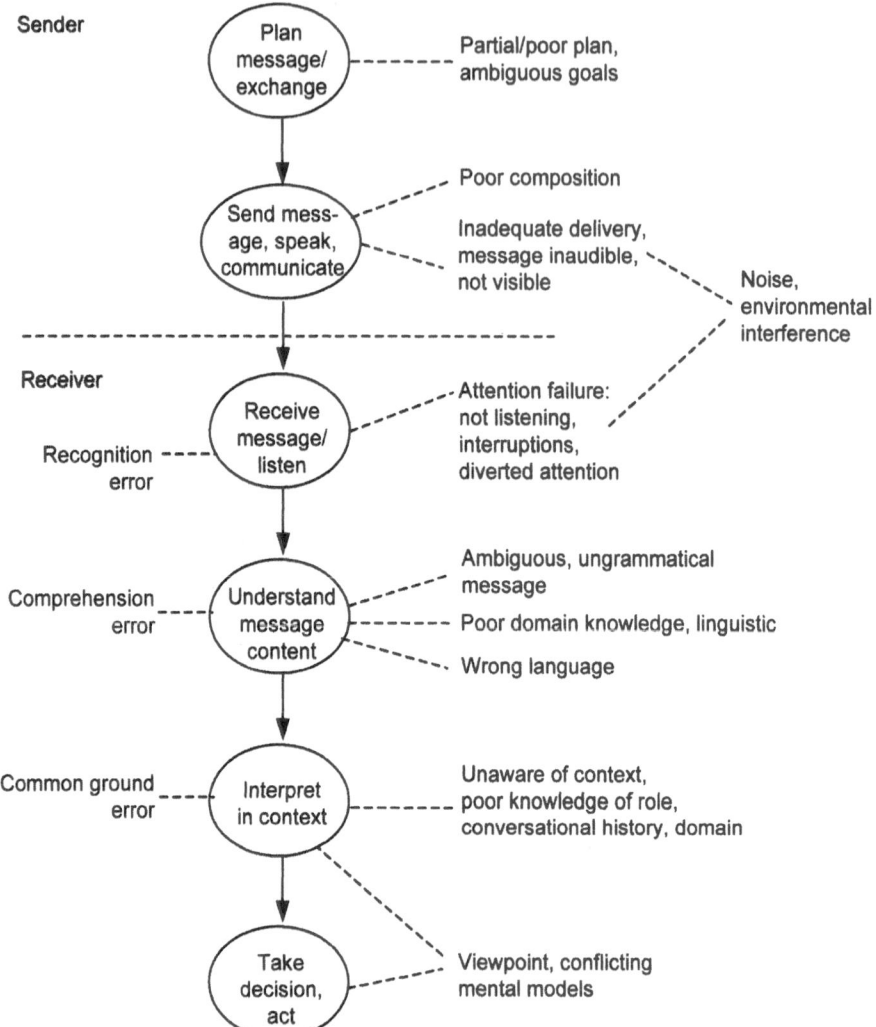

Sender

Plan message/ exchange — — — — Partial/poor plan, ambiguous goals

Send message, speak, communicate — Poor composition

Inadequate delivery, message inaudible, not visible

Noise, environmental interference

Receiver

Receive message/ listen — — Attention failure: not listening, interruptions, diverted attention

Recognition — — — — error

Comprehension — — — error

Understand message content — — Ambiguous, ungrammatical message

— — — Poor domain knowledge, linguistic

— — Wrong language

Common ground — — — error

Interpret in context — — — Unaware of context, poor knowledge of role, conversational history, domain

Take decision, act — — Viewpoint, conflicting mental models

Fig. 4.1 Cognitive model of the discourse process with associated pathologies.

erally an automatic and hence skilled process so it is prone to slip- and lapse-errors in failure to listen and poor generation. Indeed, spoken speech is full of grammatical and lexical slips. However, the semantic and pragmatic levels of understanding are prone to many of the previously described rule- and knowledge-based pathologies. For instance, we often fail to understand the implications of communication because of halo effects, confirmation biases, encysting, and mismatch errors. The communication pathologies and remedial treatments are described in Table 4.1.

Discourse, or the study or pragmatics, is the branch of linguistics that investigates how meaning is derived in context. A few examples will illustrate the problem: "I am going to the bank" can have two meanings depending on whether you know I am an angler or in need of money. Another way of disambiguating this utterance is to look at where it took place: near a river or not. "He's supporting the reds" can have at least

Table 4.1 Communication pathologies and remedial treatments.

Pathology	Remedial treatment	Task support
Inaudible message: communication not audible or corrupt	Request resend: improve communication channel, reduce ambient noise	Amplification, dedicated communication channels, noise filters
Not listening: receiver not paying attention	Request attention, test understanding	Alerting, store and repeat message
Semantic misinterpretation: words or grammatical construct not understood	Simplify language, eliminate ambiguity, reduce jargon/ acronyms, check linguistic competence of receiver	Thesauri and lexicons, parsers/ interpreters, explain symbols and diagrams, query facilities
Role confusion: interpretation mistake because intent attributed to incorrect agent	Make implications and role identification clear. Provide background knowledge of roles	Image of sender integrated with speech, non-verbal communication, role information displays
Pragmatic misinterpretation: mistake caused by failure to appreciate the location and setting of the conversation.	Provide better explanation of background and domain knowledge	Information /multimedia integration, cross-referencing, deictic interaction, history and contact information
Plan failure: utterance not generated or incomplete	Clarify, request repeat	Reuse library previous discourse
Plan failure: conversation not linked to action/doesn't make sense	Clarify viewpoint, restart conversation to establish common ground	Explanation and query facilities for context/role information

two meanings, which you will interpret according to your knowledge of the subject: is he a Liverpool supporter or a member of the Socialist Workers Party?

Language is inherently ambiguous through polysemy (one word has many meanings) and synonymy (several words refer to the same thing). Furthermore, we constantly take short cuts in speech and written language, called anaphor and ellipsis by linguists; referring to another person or object by a pronoun ("it's him"); or referring to a whole topic by a variety of constructs ("So you can appreciate, it's difficult"): something is difficult, and if we had been conversing about "discourse theory" you would know that is what is difficult.

Another problem for precise meaning is the lack of scoping in natural language. Take the utterance "Mary's mother gave her a cake on her favourite plate; she knew she would like it". Did Mary like the cake or the plate? Without knowledge of context we have no way of knowing. Meaning is often constructed by the context of a conversation. In RE we have a severe problem because little context is shared between the users and requirements analyst. The user knows about his or her domain in detail but this knowledge is not immediately available to the requirements engineer. Hence some statements will not even make sense because they depend on domain knowledge and special interpretation of jargon, or domain specific sub-languages; for instance "to strike a deal and bid an overnight" might mean little to you unless you were a foreign-exchange dealer. However, these are the easy ones to spot. More dangerous are statements which seem to make sense but have hidden interpretations which are not shared. Try, "We prioritize passengers who have been delayed from previous flights"; this sounds fine, but does it mean that delayed passengers automatically get on the next flight, or that they might if it isn't too heavily booked? So we need to be aware of context and be on guard against how different contexts can change interpretation.

4.2 Conversations and Context

Conversations take place in a context that consists of three facets: role of the participants, place where a conversation happens, and the time. To take each in turn:

The *roles* and *number* of participants determine the conversation structure and how it is managed. First there is a volume effect. More participants means either less time for each to speak, or some inequality between speaking and listening roles. Small groups of up to 4–5 people can manage a conversation with reasonable equal sharing of the "floor space" or time when they are speaking to others. Larger groups need more conversation management and a facilitator who makes sure everyone has a turn. Groups of 20 or more need more formal control of floor space by a chair and effective conversations become difficult.

Roles also involve the status and job description of the participants. For example, analysts and users will bring a set of assumptions about each other's role into the conversation, such as the users assuming the analyst just wants to sell them an expensive piece of technology and doesn't care about their problems; while the requirements analyst assumes the users are well intentioned and will tell them everything they want to know. Roles may imply power relationships which skew responses. In meetings between managers and subordinates in an organization with an authoritarian culture, a subordinate may be reticent about taking the conversational initiative and unlikely to challenge the managers' statements even if they are wrong.

The *location* in which a conversation takes place also has an important influence. If people are on their home ground they tend to be more confident and open. More formal, foreign settings intimidate people; indeed some locations are designed to do just that, such as law courts. Hence in requirements elicitation it is a good idea to go to the users' workplace and conduct interviews in their office (Beyer and Holtzblatt, 1998).

The third component of context is the *time scale* assumption. Conversations can be about the present, past or future, but frequently the reference to time may not be made clear, e.g. "we clear all outstanding payments in 60 days" might be a general policy goal which the company has always adhered to or a future intention to improve poor payment. Also the policy rule it expresses is ambiguous, because payments might be cleared when they are 60 days old or the meaning might be that all payments are cleared within 60 days, so none are older than 59 days. Conversations may also have a history, such as when we take up an argument at a point where we left off. In requirements analysis a series of interviews with a user will be a conversation which develops over time. Reminding the listener of the previous conversation helps to situate the opening of a new exchange. It also helps the other participant recall the relevant topics from long-term memory.

Conversational contexts don't have sharp boundaries. Some participants may be full members whereas others may be partial participants. Take a consultant talking to a patient in a teaching hospital. The consultant and the patient will both be involved; furthermore, the consultant may bring other trainee doctors into the conversation when appropriate. But there are other people present, such as nurses, who are not part of the conversation. They are peripheral members or eavesdroppers. However, eavesdropping has its advantages: you can monitor the conversation of others more fully because you are not participating. Eavesdroppers in RE conversations can pick up points that the primary participants missed, and help to summarize meetings.

Context is important because we create meaning by using our knowledge of the context in which a conversation takes place. In RE we want to improve the process of arriving at a commonly understood meaning, what Clark (1996) calls "common ground". Common ground, without going into the complexities of discourse theory, has two main components:

- shared belief about facts and knowledge which are referred to in conversation;
- shared externally perceivable entities to which participants in a conversation can refer.

So if I (the author) and you (the reader) were both in the same room with this book we would share some common ground in that we could both assume we understand English (though not necessarily as a first language) and that we could pick up the book and refer to a chapter or page number to focus conversation. But note that while I have attempted a text-based conversation with you across the medium of this printed page, how little common ground we actually share. I know little about you. I can assume something about your background and interests because you have bought (or borrowed) this book. But I have to guess about how much you know about RE or linguistics. If we were to meet face to face I could increase our common ground by asking questions to develop my understanding, and then tailor my response appropriately. The process of requirements analysis is not dissimilar. The analyst and domain expert start by sharing little common ground. By a process of learning, mediated through interviews, the analyst gradually increases his or her understanding. But note that common ground is a two-way process. The user or domain expert may know nothing of what the requirements analyst/system designer can do, so a true common ground between them involves several exchanges:

- from domain expert to analyst, an understanding of the domain;
- from analyst to domain expert, the potential technologies and solutions to problems;
- for both, external representations that they can understand and hence document the subject matter.

The basis for developing common ground comes from our interpretation of language by reference to memory (see Chapter 2). Interpretation unfolds according to our perception of the other person in a conversation and how much we know about them. For instance some of the factors we might take into consideration in RE meetings are:

- *Seniority*: chief executives have little time to spare and are interested only in high-level policy. Most analysts share little common ground with senior managers, so we have to learn about the company values, mission statements, strategic objectives, etc., to prepare some common ground.
- *Educational qualifications*: if we are conversing with a university graduate we can assume a level of training and occupation that is less probable for an unqualified manual worker.

- *Occupation*: we can develop common ground by understanding the job-related jargon and issues for foreign exchange dealers, doctors, motor mechanics, etc.

Memory supplies us with prior knowledge that we can apply to interpreting conversations with others, and inferring their intentions. Classification is an important means of developing common ground, so it behoves us to find out as much as we can about our users before we meet them. Another component of common ground is the conversational history between individuals. If we enter into a regular series of meetings our mutual understanding will increase by virtue of the memory we form of the other person. Such memories are fallible, so notes help.

Managing understanding is vital in RE as problems arise when understanding is not shared. Clark proposed that communication involves "action ladders" in which shared understanding is built up in stages, as follows:

- Participant A communicates an intention to B: "Let's go for a drink".
- B interprets the utterances and responds to indicate that he/she has understood: "OK, the nearest?".

However, this begs the question about how understanding is communicated. This might be either explicit acknowledgement or tacit understanding. Danger lies in the latter. For instance in the exchange above, B could:

- not acknowledge A's proposition but proceed on the assumption of mutual understanding;
- verbally reply with an understanding, e.g. "OK, fine" , but not really understand. Are they going to a pub or a café for a drink?

Common ground is not just mutually shared belief but also awareness by both parties that they do indeed have a common view. This is where common ground can often break down. We have an unfortunate tendency to assume the positive in life without seeking evidence to refute it (remember confirmation bias in Chapter 2), so in many meetings two individuals may part with completely different understandings about the same topic. Testing understanding and making beliefs explicit is vital for establishing and maintaining common ground. Communication is bound up with action and we can only judge how successful we have been when we get feedback from the other party about whether our speech has been understood and then acted upon. In RE we need to record the results of conversation, since these actions make beliefs observable and allow the other party to confirm that a common understanding is shared. Requirements and specifications record intentions to act in a design; however, observing action in a working prototype is a more powerful means of developing the link from spoken needs to action that realizes them. This leads to the next topic which is about strategies for managing conversations.

4.3 Conversation Structures

A second approach to modelling communication is to use generic patterns of discourse acts to model how the exchange of messages should be structured between people in order to achieve a shared task goal. Primitive conversation patterns were originally described by Sacks, Schegloff and Jefferson (1974), while larger patterns associated with tasks were specified in the conversation for action research by

Winograd and Flores (1986). Although natural human discourse is very variable, when associated with specific task goals such as buying and selling goods, natural regularities are observed; hence, patterns have utility for planning task-related conversations.

Conversations are made up of turns or exchanges between people. Linguists have classified different types of utterance according to their function. Some utterances are intuitive common sense, such as questions and answers, but others have more subtle functions. One of the first classifications was made by Austin (1962) who distinguished between utterances in which the speaker wants the listener to do something (termed *perlocutary* acts), compared with acts in which the speaker wants the other party to understand their meaning (termed *illocutory* acts), and the basic act of communicating information (*locutory* acts).

Illocutory acts were refined further by Searle (1969) into six categories:

1. *Assertives* aim to get the listener to acquire a belief or fact, e.g. "it will take eight weeks to implement". This category includes diagnoses, suppositions, conjectures, predictions, etc.

2. *Directives* are commands and requests for either information or action from the other party, e.g. "please record minutes of this meeting".

3. *Commissives* commit the speaker to do something, volunteer to take action, e.g. "I will find out about the development environment".

4. *Expressives* communicate feelings, apologies, congratulations, greetings, etc., e.g. "I'm sorry I misunderstood".

5. *Effectives* change the state of the world and depend on the roles of the participants, e.g. "we will have to call a halt to requirement gathering"; the effect depends on a project manager's authority.

6. *Verdictives* change the state of the world and depend on the role and institutional context where the conversation takes place, e.g. "the system does not conform to requirements" as a judgement made by the quality assurance manager.

Although discourse acts still tend to attract some controversy, the fact that classification schemes can be devised which enable independent observers to categorize conversations with 70–80% agreement is testament to the fact that we can recognize discrete meanings in conversation (Winograd and Flores, 1986). This is important for RE because if we can recognize certain patterns in conversations then we can plan ways to respond to them to make information gathering, persuasion and negotiation more effective.

Conversations should be co-operative and obey the set of principles or maxims proposed by Grice (1975) as duties that the speaker should bear in mind:

● *Maxim of quantity*: make your contribution as informative as is required but do not give more information than is required. The level of detail is hard to interpret, but for the analyst it boils down to asking questions that are relevant and at the appropriate level of detail, while for the user it is the obligation to reply with sufficient detail. However, providing extra detail is rarely harmful in RE and more mistakes arise from omissions than information overload.

● *Maxim of quality*: do not say what you believe to be false; do not say that for which you lack evidence. Both parties in RE conversations should comply with this maxim, but politics or fear of a new system can lead users to be "economical with the truth" and analysts can promise technologies that they are unable to deliver. The need for evidence is not easy to interpret, as often conjectures can help brainstorming requirements, and evidence for effective solutions can be sparse during early requirements analysis.

- *Maxim of relation*: be relevant. This is self-explanatory but can be hard to manage in meetings when individuals can wander off the point to air grievances or discuss their favourite hobby horse.
- *Maxim of manner*:
  - *Avoid obscurity of expression*: analysts and designers tend to use too much computer jargon, but user domains also have their own obscure terminology. Establish a glossary of terms and avoid jargon where possible.
  - *Avoid ambiguity*: this should be the watchword of all requirements engineers.
  - *Be brief*: verbose, rambling discourse helps nobody.
  - *Be orderly*: conversations should have a clear structure.

Grice's maxims are good guidelines for RE dialogues, but as with most principles they require interpretation in practice. More advice can be given as patterns of discourse acts that achieve a particular goal in RE tasks. It should be noted that natural conversation patterns are very variable; however, we are interested in patterns for planning requirements analysis dialogues rather than creating a theory that predicts all natural human dialogues. Like most plans in life, they will be revised in the light of circumstances as the conversation progresses.

4.4. Non-Verbal Communication

Communication involves more than just speech and language. We also communicate non-verbally by facial expressions, gesture and prosody (voice tonality). While most non-verbal communication reinforces the message of spoken communication, gesture can play an important role in controlling our focus of attention. When we point to an object it naturally focuses conversation on that object. Gesture can supplement spoken conversation in several ways, for example:

- drawing attention to an object or fact, e.g. by pointing to a rectangle on an entity relationship diagram;
- locating an object, e.g. placing an icon representing a table in the layout sketch of a building;
- tracing pathways by following arcs or pointing to objects in a sequence; tracing can also be used to demonstrate how something works, or causal connections between objects;
- expressing approval by thumbs-up gestures or head nods.

Surveying the wealth of human gesture is beyond the scope of this chapter, so the purpose here is to point out that non-verbal communication can supplement discourse acts or sometimes constitute an act on its own. We also communicate by other media such as image, text and artefacts in the world. Some signs have meanings that form part of our common sense knowledge, such as interpreting a compass needle's direction as pointing north. However, designed symbols are different and these have to be learned, e.g. the node and arcs conventions on entity relationships diagrams.

4.5 Dialogue Acts and Patterns

Conversations are built up of pairs of *dialogue acts*, the most basic of which is the question and answer. Adjacency pairs lead to *dialogue patterns* called *exchanges* that are motivated by a shared goal. Exchanges are typically composed of three acts, but longer chains are possible. Some examples are:

● question – answer – evaluate
● check (test question) – answer – confirm/deny
● propose – agree/disagree.

A set of acts for requirements-oriented dialogues is as follows:

● *Request*: the speaker requests information from another. This category includes explicit and implicit questions seeking information on a topic, e.g. a user's goal, how an event occurs, or an example of an object, event or problem. Examples of explicit questions are: "What are the main problems which arise when retrieving staff information?" and "Can you tell me how retrieval takes place?"; and an implicit request "Talk me through the invoicing process" . Requests may also be for a justification or rationale for a topic under discussion, and the causes and effects of situations; e.g. "Why is that so?" and "Is this because updates are batched?".

● *Inform*: the speaker provides information, either unprompted or answers questions, elaborates details, and gives examples; e.g.: "The main problem is response time" and "Retrieval is partly manual and partly automated". Also included are giving background information, causes, results and examples which demonstrate rationale, e.g. "The problem arises because of the system's response time".

● *Check*: the speaker requests clarification of a current topic, e.g. "So is it delivered before or after the dispatch note?". This act tests one's understanding after information has been received. It can be manifest by either explicit questions or non-interrogative statements, "So first you go through the pre-flight check list then you get OK to start engines". Implicit checks are often signalled by use of voice tone, or *prosody*, to let the other party know that there is some doubt about the facts.

● *Confirm*: the speaker confirms his/her own or someone else's understanding of a current topic, usually in response to a check; e.g. "Yes, it's the goods that are delivered after...".

● *Summarize*: gives a synopsis of a preceding conversation, an overview of topics which have been discussed.

● *Propose*: the speaker proposes an action or topic to be discussed, or makes statements about planning action or conducting a meeting, e.g. "Let's write these requirements on the whiteboard". Proposals may be a solution to a problem, stating a requirement, or suggesting how a problem may be solved.

● *Augment*: to add new facts and propositions to develop a proposed solution, course of action or refine a requirement.

● *Agree*: the speaker confirms that he/she shares the same belief in a proposition and course of action and signals that they support a proposal.

● *Disagree*: the speaker signals that he/she does not share the same belief as a previous speaker and would not support a proposal.

- *Correct*: state why a previous assertion was wrong and explain why, giving an alternative proposal or assertion.
- *Record*: edit or create a text or diagram to record the facts and topics which have been discussed.
- *Demonstrate*: operate a computer system or other machine to show how it functions. Human actions and tasks may also be demonstrated.
- *Explain*: inform the listener about some issue with additional information so they understand how something works, what caused an event to happen, etc.

Discourse acts can be built up into patterns of exchanges which are linked to a goal held by one or both of the participants in a conversation. In many conversations, RE included, the goal is agreement on proposals, action or a set of beliefs that constitute the system requirements.

Longer patterns are composed from exchanges which can be embedded within each other or chained together. RE conversations have goals that are set by the tasks, so dialogue patterns are structured to provide the means to achieve these ends. Requirements tasks imply the need for acquiring information, agreeing proposals and checking that proposals are understood by all parties. For instance, requirements elicitation and acquisition involves seeking information from users, while negotiation will need patterns to manage agreement between parties. Analysis and modelling need patterns to refine requirements and explore their implications. Finally, requirements validation has to check that users understand and agree with the stated requirements. The following sections describe dialogue patterns and conversation management techniques for each task in turn.

4.5.1 Requirements Elicitation

This task is carried out by a variety of techniques including observation, documentation analysis, structured and unstructured interviews. The analyst has to capture basic information about the domain: who the important agents are, what they do, what is the purpose of the system, who are the customers and users, what it produces, where it is located, and so on. The objective is to capture lists of facts to start modelling:

- *agents and objects*: who does what, what things are involved;
- *attributes and properties* of the system and objects;
- *components*: what are the major parts of the existing system, sub-systems;
- *functions*: what does the system do, major processes or activities;
- *context*: what or with whom does the system communicate, where are the boundaries, where is it located.

Closed questions which require a yes/no answer are generally not effective because the analyst makes assumptions about the subject matter. Semi-closed questions are effective for following up points, whereas more open questions are better for intention capture. Questions also have to be targeted at the appropriate people. The following advice may be familiar from systems analysis textbooks but is worth reiterating:

- Ask senior management questions about high-level aims and policy; reasons why the system is important; and critical success factors (CSF). Senior management time is always at a premium so a list of questions for a structured interview is essential. Senior managers may either have revolutionary visions for change and can describe scenarios for new systems in some detail, or they may just set policies and targets which the designer has to interpret.

- Target questions about goals, objectives, performance criteria, non-functional requirements and outlines of tasks on middle-level managers. These managers will consume the output of the system but are less likely to directly operate it. Middle managers are more likely to perceive problems with the current system and come up with suggestions for improvements.

- Questions about details of tasks, procedures, events, actions, objects, etc. are best reserved for operational personnel who will actually use the system. Beware that operational personnel's view of how the system should be designed may be coloured by the way the current (legacy) system operates; however, most requirements and user suggestions for improvements will be useful.

First a set of general scoping questions are used in the first stage in requirements investigation. The boundaries of the investigation need to be discovered early in requirements analysis to set the terms of reference for subsequent analysis and modelling. The following scoping questions are useful probes to establish system boundaries:

- *What problems does the system need to solve?* The answer identifies problem-oriented requirements, subsystems and high-level functions. It may also elicit user dissatisfaction with the current system and indicate areas which need re-designing and improvement.

- *What problems could the new system create?* This is a useful reality check to make sure the new system does not upset too many useful existing processes and hence cause more harm than it is worth.

- *What processes will introducing this system change?* This helps to focus on the socio-technical impact of the new system and the changes it may cause in existing processes. Many processes need to be changed, but that might not be a bad outcome if previous processes were inefficient.

- *Who will own or control the new system?* This question is useful for uncovering the politics behind an organization and teasing apart primary, secondary and tertiary stakeholders. The question will probably require supplementary explanation and follow-up questions to find out who is responsible as the budget holder and initiator of the system, as well as identifying the users and operators. If several owners are identified this should ring alarm bells about possible conflicts in interest.

- *What environments will the system be placed in?* This uncovers the possibilities that the system could either be mobile or distributed internationally. Follow-up questions need to probe cultural assumptions, the range of languages to be used, and where the system will be located. Supplementary questions may explore the range of temperature, humidity, noise, dust/dirt, vibration, etc. which may be encountered, whether the system will be moved or is mobile. Answers to these questions feed into the constraints on design, maintenance and robustness, non-functional requirements.

- *What precision is required for the system?* This question will elicit requirements for accuracy of output as well as functional performance. Explanation will be necessary to establish what precision means in the domain, i.e. precise calculation (operational control tolerances within 1 mm), operational reliability (mean time between failure of 10^{-4} days), etc. Precision questions are necessary for many non-functional requirements and performance benchmarks, as well as discovering the necessary and sufficient requirements by which the system can be judged.

- *How much is this system worth?* This question elicits opinions about how much the client values the required system. Explanation in this case will be necessary for how values are to be estimated, the cost of replacement, cost of loss, or the penalty of not having the system. Value estimates will differ between the customers and managers who commission the system and the users who will operate it. The former may answer in terms of the budget they have allocated for the development whereas users will see worth in terms of saving time, effort and avoiding mistakes.

Questions need to be ordered in conversational structures to check detail and follow up important points. For this purpose, dialogue patterns are described that propose a typical exchange between the analyst and user. Exchange structures for requirements conversations are suggested as "tools for thought" for developers to rehearse their explanations and practise questioning skills.

The pattern for elicitation is composed of general questions, answers and follow-up questions. The main conversation pattern is interviewing. Two conversation roles are present: the interviewer (requirements engineer) and the interviewee (one or more stakeholders). The conversation consists of exchanges between the interviewer seeking information, and the interviewee giving information, possibly followed by clarification of replies.

A typical pattern for requirements elicitation is illustrated in Figure 4.2.

Questions are initiated by the requirements analyst, followed by answers and a cycle of checking and clarifying facts. Facts are elicited by semi-closed questions which contain some information indicating the expected answer, followed by check-

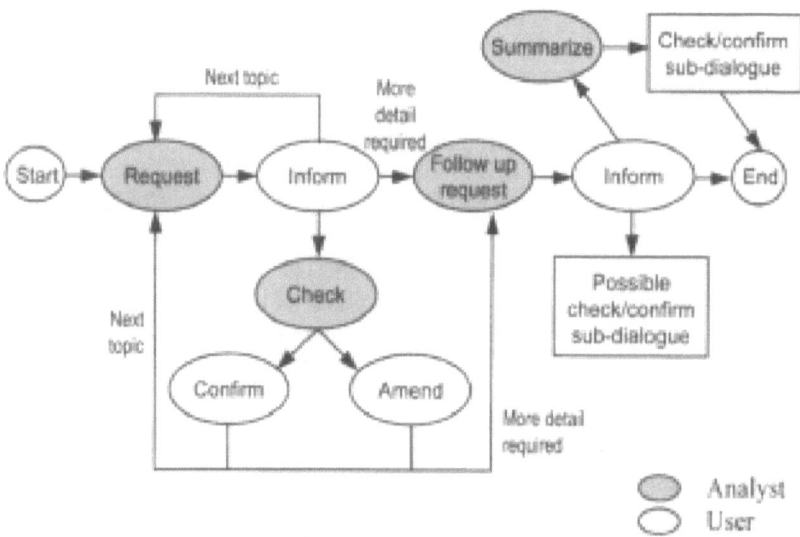

Fig. 4.2 Dialogue pattern for fact-gathering interviews.

ing statements which test a user's belief (e.g. "So in normal operation distribute the hazardous cargo in different parts of the ship") and use of gesture to refer to external representations. The pattern finishes by summarization to check that the analyst has acquired all the relevant facts and understood them correctly. Representation support consists of lists, models, informal diagrams and sketches. This basic pattern can be specialized, as follows, according to the type of facts that are being acquired.

Intention Capture Discourse

This elicits user goals, aims, objectives and so on, which will be translated into requirements. The structure of this pattern (Figure 4.3) encourages recursive drill-down requests to explore goals at increasing levels of detail. The pattern is geared to goal analysis that creates a goal hierarchy (Loucopoulos and Karakostas, 1995; Bubenko, 1993) and the relationships between goals with the following questions:

- *Is this goal composed of sub-goals?* This indicates decomposition to create two or more "child" goals that contribute to the parent goal.
- *Does this goal depend of completion of another goal?* Follow-up questions probe whether the parent goal needs all its sub-goals to be completed, or just some of them. The answers indicate AND-OR branches in goal trees, and give early hints about process dependency.
- *Will this goal inhibit completion of another goal?* This indicates clashes between goals which need to be resolved. Inhibition may be either complete, i.e. both goals cannot co-exist; or partial, in which one goal impairs but does not prevent another.

Questions about who owns or is responsible for the goal are a useful check on scoping the system boundary to test for goals that are outside the scope of the current requirements model. Open questions are asked, e.g. "Are there further information needs?", and scenarios may be used to critique the scope of the system, e.g. elicit a scenario of problems and failures in the current system and discover goals for new functions.

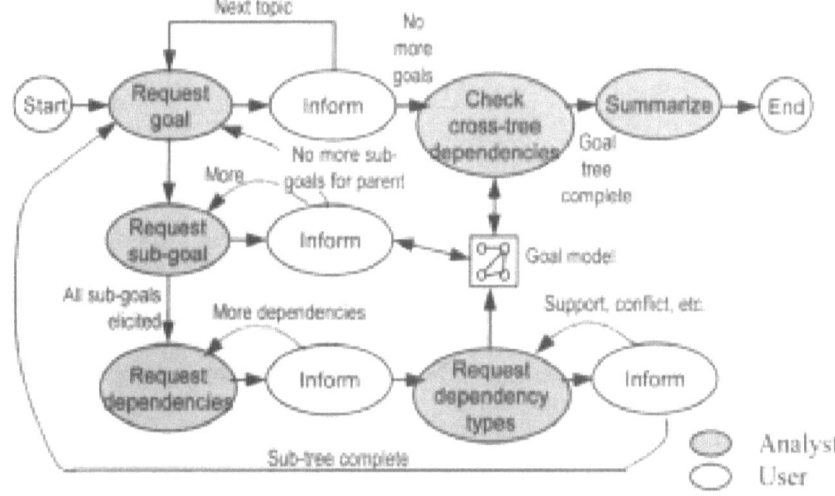

Fig. 4.3 Dialogue pattern for intention capture discourse.

Higher-level goals may be referred to as policies, aims and objectives. These are all intentions but we cannot say how they will be achieved directly. For instance, a high-level aim might be "to increase customer satisfaction". This intention can only be understood in terms of lower-level goals which unpack the aim as "to reduce processing time for complaints", "to improve personal service in ordering" and "to reduce lead times for delivery". Low-level goals can then be operationalized in terms of how they might be carried out. Goals are statements of intention, but they are also indications of how we want the world to be if the goal is fulfilled. This leads to the next pattern: procedure capture.

Procedure Capture

This pattern uses "how" questions to elicit details of the tasks, processes or activities that lead to a goal being achieved. Procedures specify how goals are achieved and start with an existing state (the pre-condition), followed by a sequence of actions that change objects within the system to arrive at the goal or end state (post-conditions). Procedure capture dialogues are based on different types of goal (Anton and Potts, 1998; Potts *et al.*, 1994; Sutcliffe and Maiden, 1993a):

● *Achievement goals*: the procedure causes some change in the world which results in a new state specified by the goal. Achievement goals assume that the current state is unsatisfactory and specify what must be done to improve unsatisfactory states, e.g. "All damaged goods must be replaced within 24 hours".

● *Maintenance goals*: in this case the state is already satisfactory and the system's duty is to preserve it. Maintenance goals have to detect deviations in system states and take corrective action, e.g. "If the temperature in the engine exceeds 180°C then open the coolant valve until temperature is decreased below 180°C".

Further goal types can be proposed, such as maintaining proportions in a population (Sutcliffe and Maiden, 1993a) and ceasing an operation (Van Lamsweerde and Letier, 2000); however, most fall into the achievement/maintenance categories. The pattern for eliciting procedures for these goals is given in Figure 4.4.

The outcome of this dialogue is a record of actions in structured sequence, that achieve the state or post-conditions that satisfy completion of a goal.

Event Capture Dialogue

Event-driven analysis is a complementary approach to goal-driven requirements acquisition. Event analysis focuses on the inputs and output from the system and asks two simple questions:

● *What input events must the system respond to?* Input events may come from users, other systems or the external world. Follow-up questions then probe the source of the event and which component in the system is responsible for responding to the event. This leads to identification of goals or functions.

● *What events or output does the system need to produce?* Output events may be messages, reports, commands and actions that are communicated to users, other systems or agents in the external world. Follow-up questions probe the destina-

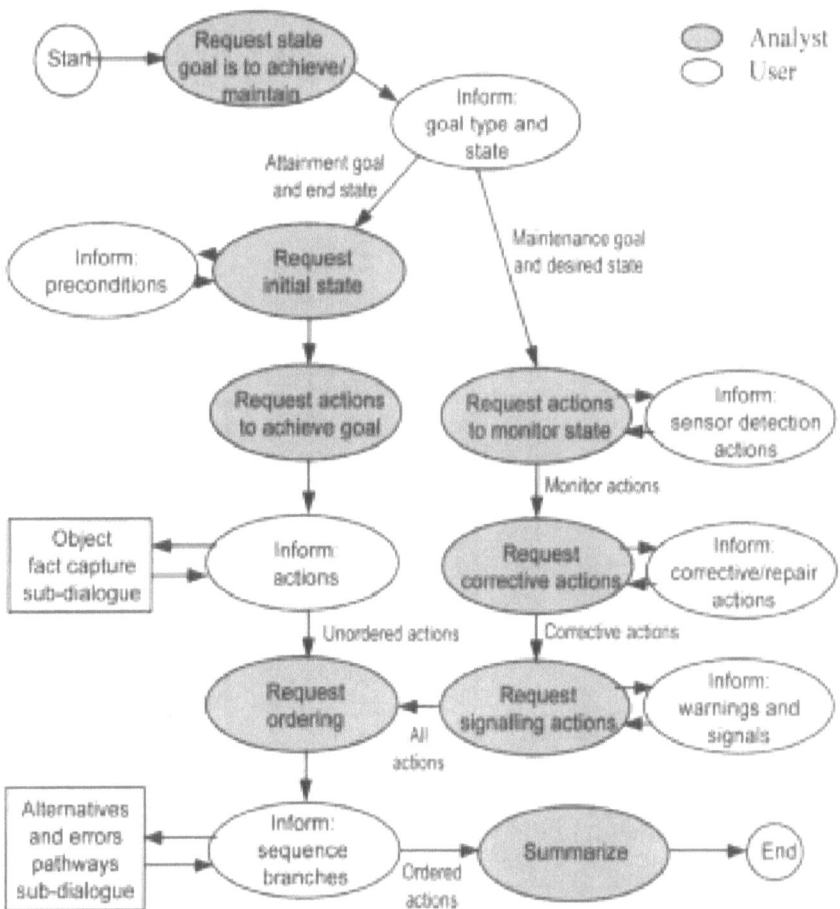

Fig. 4.4 Procedure capture dialogue.

tion of the output, the nature of the output content, and which system component is responsible for producing it.

Questions focus on input and output events across the boundary of the intended system. User input is cross-checked against the requirements specification. If no system function exists to deal with the particular event then a goal must be added to the requirements specification. The pattern for inbound and outbound event analysis is given in Figures 4.5 and 4.6.

For inbound validation, the impact of missing, mis-timed, and inappropriate events may be analyzed following an entity life history approach (Jackson, 1983) to suggest system requirements to deal with omissions, commissions, unexpected events, excessive volumes of input and dangerous events in certain contexts. Entity life histories are useful for drafting requirements specifications because they explicitly represent event sequences and dependencies. Questions encourage the users to explain the range of different event types which might be input, changes in event attributes, likelihood of delayed, missing events, etc.; for example, "What happens if the emergency warning is delayed?". When the developer discovers events which the current specification cannot deal with, follow-up questions elicit what the system should, or should not, do in this circumstance.

For outbound validation the focus is on the acceptability and impact of system output on users. A scenario of the external world can be used to test the acceptability of output, and the users are prompted to imagine permutations on this scene, and more importantly the effect that different environmental contexts (e.g. climate, location) and user roles may have on the acceptability and appropriateness of the system output. Questions focus on whether the information would be appropriate

Fig. 4.5 Dialogue pattern for event fact capture for inbound events.

or acceptable to different user groups, and if the implication of system decisions will be workable in the real world, e.g. "Will detailed instructions be understood by the emergency crews?", "Can warnings be heard in all compartments?". Obstacle analysis from the inquiry cycle (Potts *et al.*, 1994) can help to frame questions for validating system output.

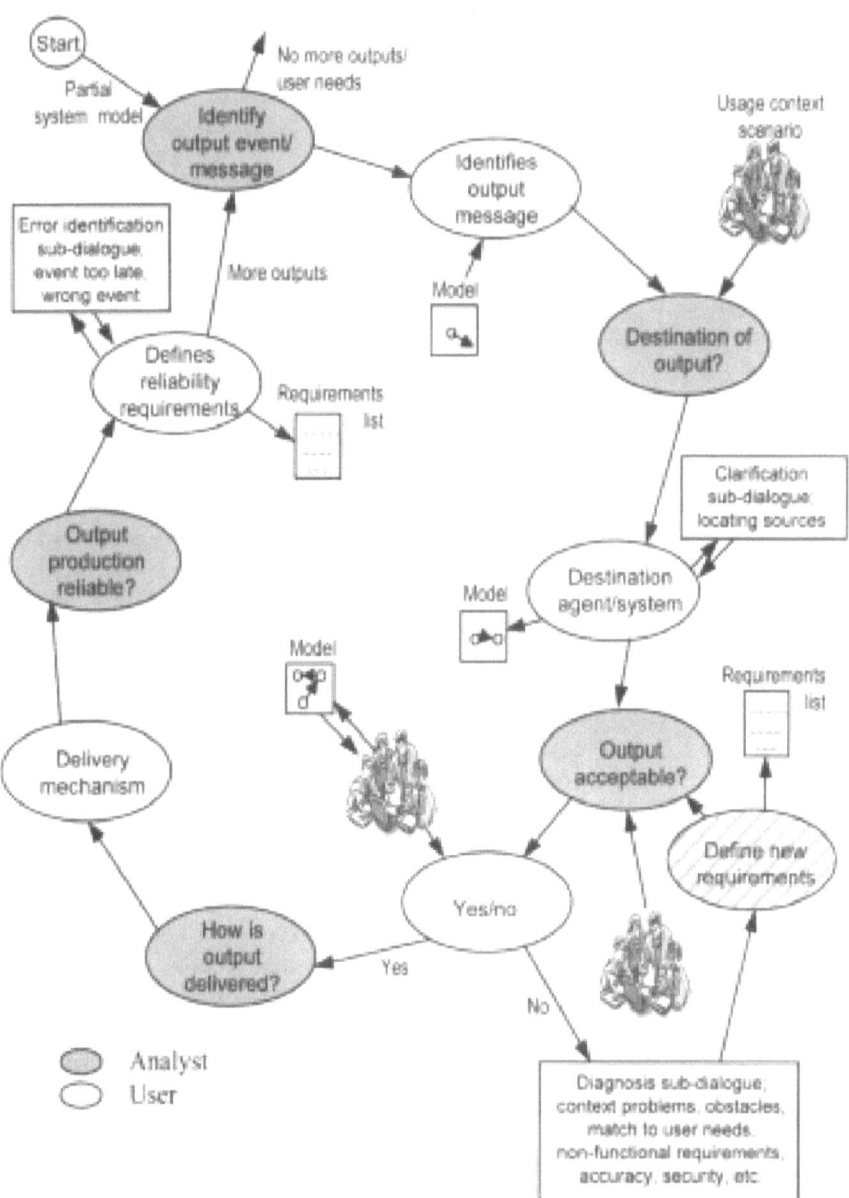

Fig. 4.6 Dialogue pattern for event fact capture for outbound events.

4.5.2 Analysis and Modelling Dialogues

Analysis explores the relationships, dependencies and associations between facts to create models. While there is no hard and fast distinction between requirements acquisition and analysis, and both dialogues may be interleaved within an interview, analysis patterns will become more frequent in follow-up interviews when preliminary models have been created. Analysis patterns are intended to stimulate understanding of the domain and existing system as well as synthesizing new ideas about how the future system might address goals and problems that have already been elicited.

Causal Analysis

Causal analysis may be necessary for understanding how a domain behaves or how a design might operate. Causal analysis is closely related to procedure acquisition dialogue but differs in that the explanation is incomplete or not known beforehand. The dialogue pattern starts by eliciting the location and detail of the observed problem, then suggests hypotheses about why the event occurs, followed by testing that the selected explanation can account for the known facts. See Figure 4.7.

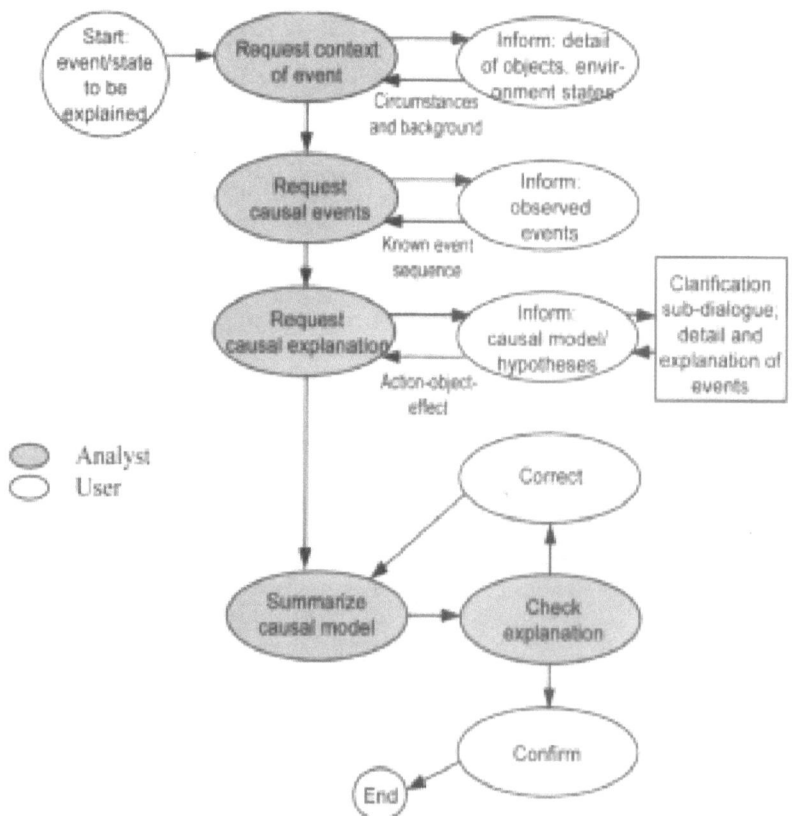

Fig. 4.7 Causal analysis dialogue pattern.

Establishing Connections and Dependencies

The connectivity analysis pattern (see Figure 4.8) investigates links between objects/agents/tasks, etc. that are then represented in a model. The links may be represented in many different notations, for instance, information flows in DFDs, functional dependencies in an entity relationship diagram, or sequential links in a chain of state transitions. The key facts to be acquired are the source and destination of the link, the type of link and then other details which depend on the modelling convention being followed, e.g. cardinality, optionality of the relationship. Tracing connections is an important activity in management of requirements to specify traceability relationships in RE management tools.

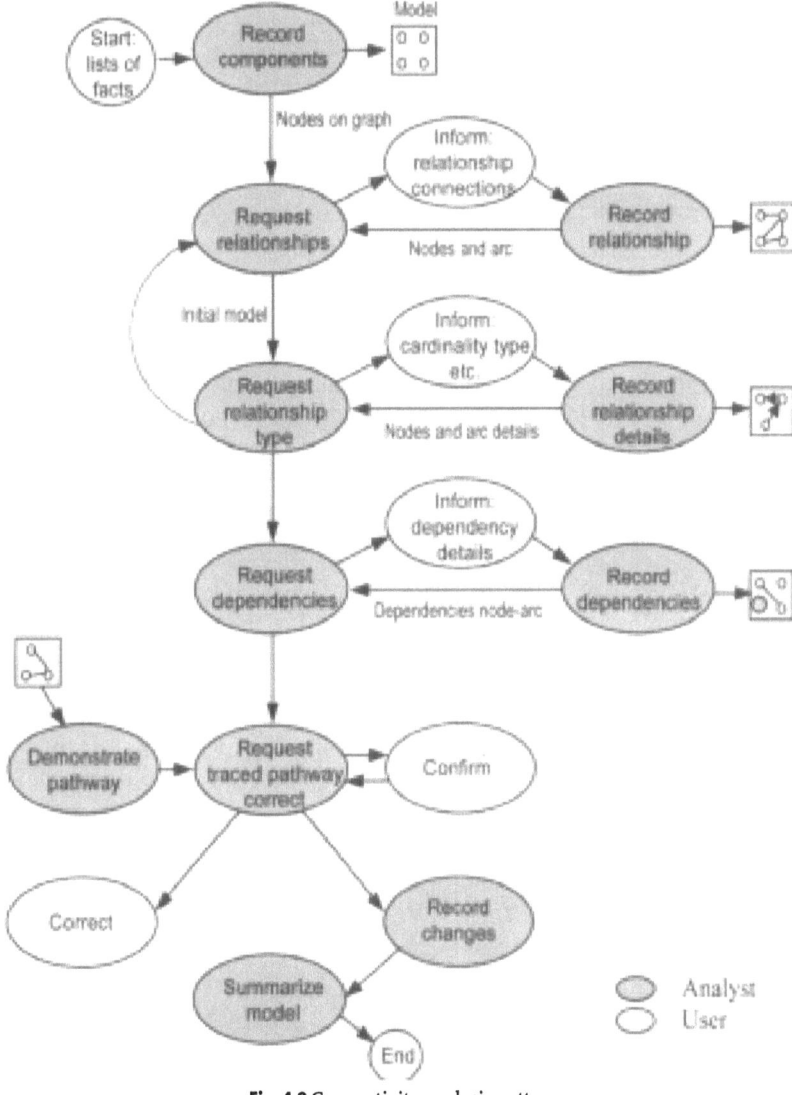

Fig. 4.8 Connectivity analysis pattern.

Dependency analysis is a specialization of the connectivity pattern that involves assessing relationships where associated objects depend on each other; for instance, where the operation of one agent/task depends on some service or resource being supplied by another, while stakeholders provide clarification or additional details. Typical dependencies in RE models are:

- *goal–goal*: one goal depends on another goal for its successful completion;
- *goal–agent*: successful completion of a goal depends on a particular agent;
- *goal–task, or goal–agent/task*: successful completion depends on an agent or task;
- *task/agent–resource*: an agent needs a resource to carry out a task;
- *agent–agent*: one agent needs the co-operation of another.

Dependencies are checked between model components; for instance, that an agent has been allocated a task, or that the pre-conditions for some system activity are met. Alternatively, dependencies may be traced between representations or along information flows and event sequences to ascertain that a sequence is correct and event pre-conditions can be met. Representation support for tracing requires relationship direction to be portrayed as well as the identity of nodes. This is necessary when the dependencies of agent/object involvement have to be traced. See i* (Mylopoulos, Chung and Yu, 1999).

4.5.3 Refining Requirements

These dialogues are synthetic rather than analytic and aim to add new components to requirements specifications. The pattern starts with a requirements model and develops new suggestions for improving the system and solving problems (Figure 4.9).

The key points for the developer are to explain the requirements issues, then propose a design solution and illustrate it with a storyboard or demonstration. The users are invited to contribute their own design proposals for the requirements or to build on the developer's proposal. The pathway followed depends on how forthcoming the users are with suggestions. Different options may be explained and justified using design rationale to compare options against design criteria, possibly leading into a negotiation pattern. However, the main goal of the refine requirements pattern is to create suggestions for design, build on those suggestions, and then agree them with the users.

This pattern may also be used in brainstorming meetings in which new requirements are elicited from users early in a systems investigation, as well as in sessions to develop new requirements in collaboration with users later in the process. Stakeholders make suggestions, build on suggestions by others, and critique each other's ideas. Representation support consists of lists for high-level requirements, sketches or storyboards of the system vision, post-it notes and other means of annotation for priorities and recording authorship.

4.5.4 Validation Dialogues

Validation involves testing that the recorded requirements are understood by users and fulfil their wishes. Hence the dialogue pattern aims to test understanding and then elicit agreement from the users to check that the requirement meets their needs and reflects an accurate statement of what is needed and is achievable. Validation dialogues assume that a model or requirements specification exists, and systematically walk through a model or list checking each requirement and associated details. Three roles are usually employed: a facilitator (session leader) who explains the requirements specification by progressing through it step by step; peers, who may be designers like the facilitator, or users; and a moderator who controls the discussion. The pattern (Figure 4.10) consists of explanations of each step, critiques by the peers, and clarification by the session leader that may then trigger a sub-discussion leading to agreement or disagreement about the specification step in question. The outcome is a critiqued specification and/or an action list for steps to be changed. Representation support consists of models that facilitate pathway analysis and notations, which allow tracing between requirements, model components and design options.

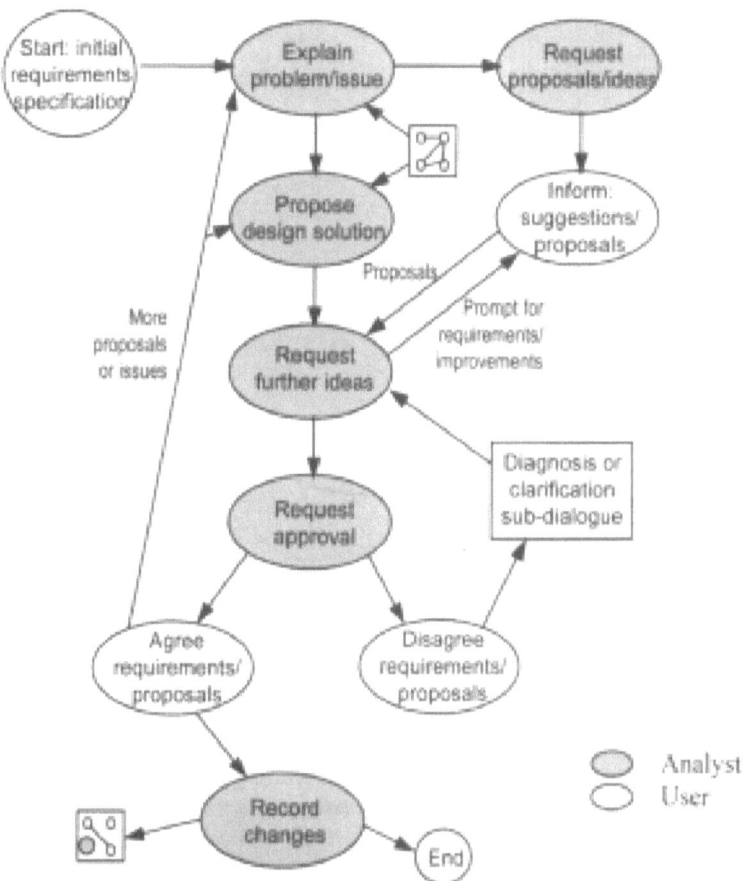

Fig. 4.9 Dialogue pattern for suggesting requirements.

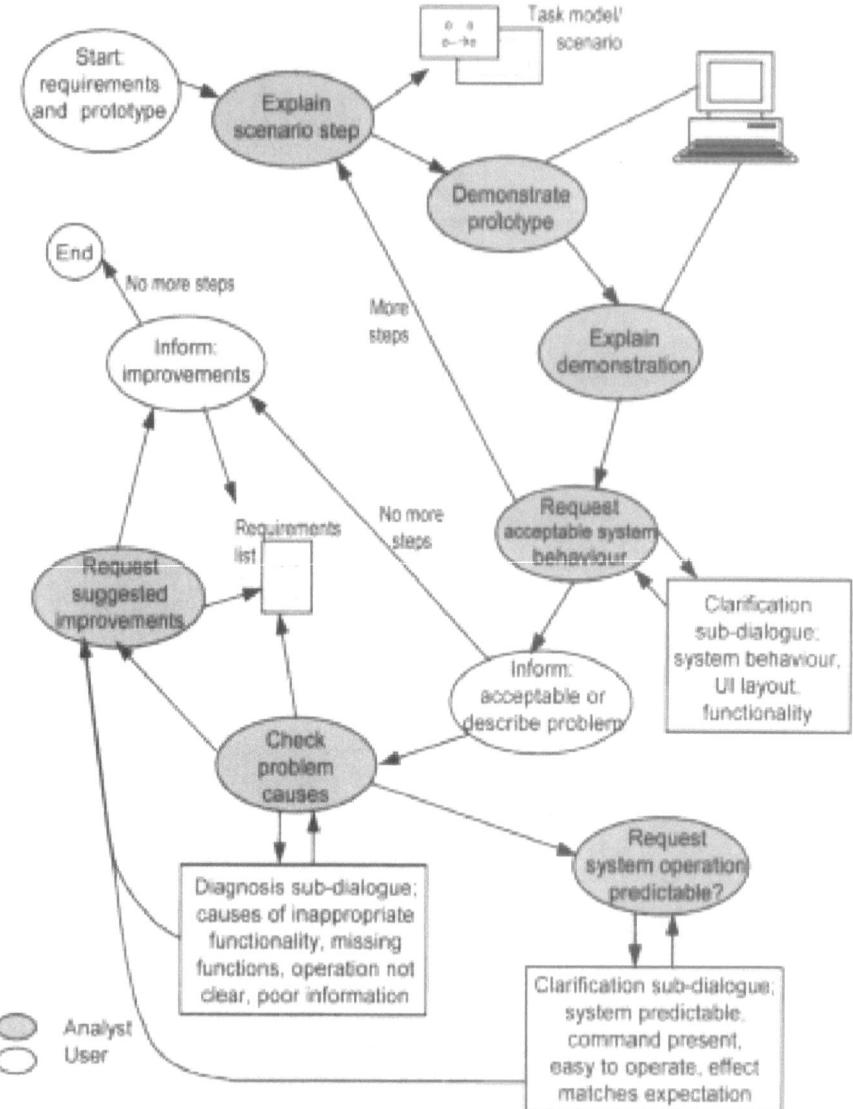

Fig. 4.10 Validation dialogue pattern.

4.5.5 Negotiation Dialogues

Negotiation dialogues involve explaining the arguments for a decision and any available options; and then exploring the advantages and disadvantages for each option in order to reach a decision which is agreed by all the users. Negotiation is frequently embedded in other RE tasks. Representation support includes lists and models to support giving information, clarifying positions and alternatives to help negotiation by making the choices visible and shareable by all parties. Matrices and decision tables help by making the interaction between choice and decision criteria clear; ranked lists and voting systems can help represent priorities. Viewpoint representations (Sommerville and Kotonya, 1998) also help to make the positions of different participants clear. In some domains a formal model of the decision can be implemented in software, so automated reasoning can be applied. Even if the decision is not automated, the act of formalizing the model often clarifies issues (Mylopoulos *et al.*, 1992; Jacobs and Kethers, 1994).

Negotiation may take several forms, so the pattern needs to be tailored for the goals and the decisions that have to be made, for example:

- *Single choice decisions*: in this case the negotiation is about whether to proceed with a course of action, or approve a design. The outcome is approval or rejection.
- *Multiple choice decisions*: these are more frequent in RE with negotiation over trade-off choices, selection between design options, and reconciling different stakeholder views.
- *Prioritization*: these ordering decisions involve negotiating priorities and the criteria by which priorities are determined.

The key to successful negotiation is making the facts clear, establishing criteria by which judgements can be made, and making sure agreements are by consensus wherever possible. In conversation management, the facilitator should give each stakeholder the opportunity to present their case. When disagreement arises the important point is to diagnose the cause. Disagreements frequently arise from hidden assumptions and dependencies, or applying different criteria to the decision, hence the party who disagrees should be asked to justify their position and make their reasons explicit.

The pattern starts (Figure 4.11) with the analyst explaining the choices to be made, the options or positions that form the subject matter for the negotiation, and the judgement criteria. A preparatory dialogue may be necessary to agree the judgement criteria; and structuring options against the criteria so that all parties gain something (the win–win condition). This smoothes the way for a successful agreement by consensus (Boehm *et al.*, 1994). If everyone can come away from a negotiation feeling that they have gained something, even if they have lost on another topic, then the agreement is more likely to last. Structuring negotiations to achieve win–win positions is of course not easy, and skilled negotiators are necessary to achieve agreement in difficult cases. In RE, stakeholder analysis is a useful preparatory activity because it prepares the initial positions and gives the analyst a means of finding potential wins for each stakeholder group.

Once the initial positions have been explained, the pattern advocates discussion of the options before soliciting agreement. This gives all parties the chance to voice opinions which might not have been captured beforehand. If agreement is reached, the negotiation is over; otherwise, the reasons for disagreement have to be diagnosed, then a new set of options and criteria built which give all the parties some gain before inviting agreement again. If exploring alternatives or relaxing the decision criteria don't work, then it is advisable to postpone the decision pending further investigation.

This concludes the dialogue patterns for RE tasks. These templates are specialized for different settings, so their use is via practice rather than being a model which is followed verbatim. The patterns provide a starting point for organizing RE dialogues and a set of heuristics which will become part of the analyst's tacit skill.

Fig. 4.11 Negotiation dialogue pattern.

4.6 Summary

Communication by natural language and diagrams is a key part of RE, but it is prone to ambiguity, inaccuracy and misunderstanding. RE aims to establish a common ground of shared understanding between users and software engineers. Common ground is built up by testing assumptions and beliefs as well as understanding the viewpoints of others. Conversations can be broken down into discourse acts. Discourse acts can be combined into patterns that describe typical conversation structures for different RE tasks. Dialogue patterns are described for fact capture, goal modelling, event-driven requirements capture, as well as causal analysis and negotiation. Each pattern specifies the acts performed by different roles – for example, analyst, user, facilitator, etc. – and how the questions, answers and proposals are linked to representations that support the dialogue.

5 Representing the Problem

The previous chapter discussed dialogues for requirements, and processes for discovering them. This chapter focuses on how requirements are represented and recorded. Representation and requirements dialogues are, of course, closely linked. Representations become part of the conversation when we refer to and point at specifications and models during the process of analysis. Representations have been a somewhat neglected part of RE, although this problem has received increasing attention in several literatures (Muller, 1991). Requirements tend to be documented as lists of things that need to be implemented, such as functions or goals. However, requirements may be expressed in models as requirements specifications rather than just lists; alternatively, requirements may be embedded in a prototype design. Clearly lists, specifications, designs and prototypes all represent requirements in some manner; the question is, what are the merits of different types of requirements expression? This chapter addresses this question and looks at how different representations fit within the RE process.

5.1 Representation Criteria

Before launching into representations, it might be helpful to ask the question: what is the purpose of requirements representations, and what do we need them to do? At a basic level, representations document and record requirements. However, we want any form of expression to do more that that. It should enable or facilitate the process; in other words, a representation should make some aspects of the world clear so that people can reason about them. How we reason depends on the RE task, so representations should be matched to the task(s) they support. For example, lists are good representations for prioritizing requirements because they can be readily sorted and ranked, whereas requirements expressed in models are more difficult to prioritize.

Representations need to be more than suitable for a task. Ideally they should also be easy to learn and use, have precise meanings, be easy to change, and so on. They need to obey a set of "cognitive dimensions" (Green and Petre, 1996) that specify their usability and comprehensibility. Unfortunately some of these properties will clash. For instance, precise meanings are found in many formal specification languages (e.g. Z: Spivey, 1989), but they are not very easy to learn. A set of general desiderata for representations is as follows:

- *Task affordances*: the concept of affordance is taken from Norman (1988), who borrowed it from Gibson (1986). An affordance is a design feature that suggests some possible action (e.g. a door handle suggests either turning or pulling depending on its shape). Here "affordance" is used to describe how the form of a representation supports a user's task.

● *Ease of learning*: a representation should be comprehensible and its notation easy to learn. This will be a property of how complex and well designed the symbol set or modelling language is. Some representations can be considered to be natural and require little or no learning because they are part of our everyday experience (e.g. hierarchy diagrams as organization charts); others are designed conventions, or languages which have a lexicon (set of symbols) and grammar (composition rules) that have to be learned (e.g. statecharts: Harel, 1987).

● *Ease of manipulation*: this expresses how easy it is to use, in order to create a model or specification, and how easy it is to change. This will be a function of complexity and the coupling between symbols in the language. Generally representations with many symbols, complex grammars and fussy constraints on physical layout will be hard to manipulate (Green and Petre, 1996).

● *Scope/coverage*: ideally a representation should be able to express a large number of different facts about the modelled world and to show the links between them. However, better coverage pays a penalty in increased complexity and decreased comprehensibility. For instance, if we represent data structures, event sequences, goals and agents on one diagram it will become a complex tangle of arcs and nodes, even though it covers many different aspects of a requirements problem.

Representations may be passive, active or interactive. Most diagrams, models and specifications are passive, whether we view them on a VDU or read them on paper. Prototypes on the other hand are interactive, as you can try them out. The reason why this distinction is important is a consequence of how we reason with, and learn, representations in various forms. RE is a learning process; hence we want representations to help us discover new facts and insights into the problem in hand. The problem is similar to the way that technology and artefacts can be designed to help students learn. We learn more effectively by doing (Papert, 1980). We also learn by discussing problems with others, a phenomenon called "learning conversations" by educational psychologists (Laurillard, 1993). Active engagement in problem solving encourages the development of rich memory schema. Passive representations don't help active problem solving, because you cannot directly interact with the artefact. Of course you can trace a pathway on a diagram, annotate or doodle on it, but essentially you have to expend the effort in doing so; the artefact doesn't help. On the other hand a prototype invites you to try it out, so you have to guess what to do (especially in poorly designed, obscure user interfaces). Once you have taken the plunge and tried a command, you are faced with the problem of figuring out whether it worked the way you expected it to. This is active engagement in problem solving, since the prototype gives you no choice but to problem solve. Whereas a cursory glance at a specification might enable you to satisfy yourself that it will work, interacting with a prototype makes you take that decision actively.

Interactive prototypes and RE support environments have much to learn from constructivist learning theory (Papert, 1980; Carroll, 1995). Taken to an extreme, there is an argument for developing RE software environments that are composed of interactive prototypes, functions such as critics to explain when something might not work, and facilities to explain reusable good design ideas gathered from experts in the domain. Fischer and his colleagues have worked on such DODEs (domain-oriented design environments) for a number of years (e.g. Fischer *et al.*, 1992). A sample environment, from NetDE (Fischer, 1996), is illustrated in Figure 5.1.

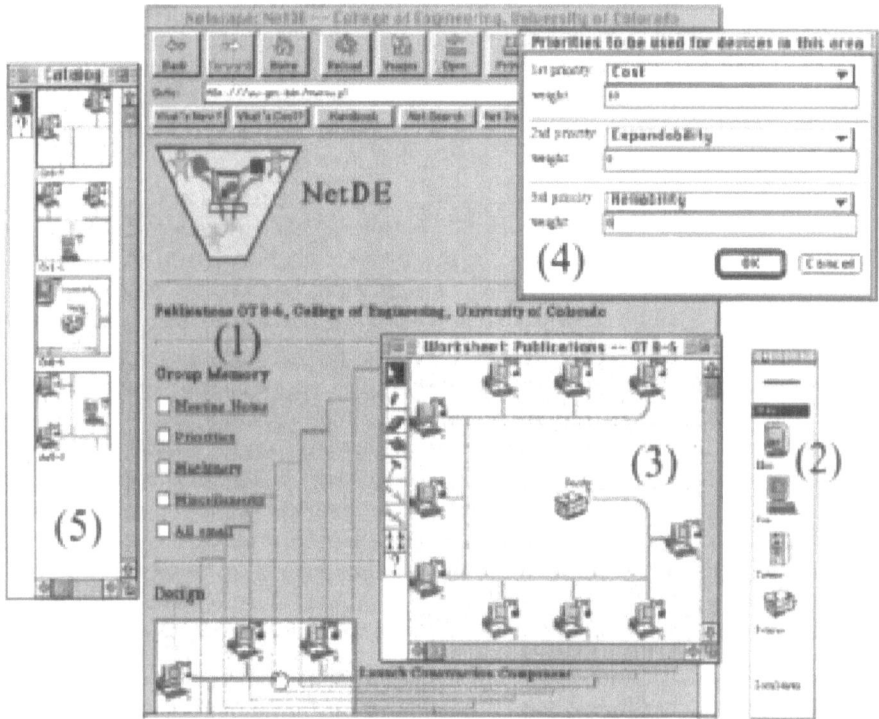

Fig. 5.1 DODE environment for Network design. The *number* refers to components of the DODE (1) information notepad, (2) palette of network design objects, (3) design workspace into which objects can be dragged and dropped, (4) critic/design advisor, (5) template in design library.

The environment contains a construction kit of components that a designer would need in the domain (e.g. a set of kitchen furniture and appliances), rules that govern composition of design in the domain (i.e. spatial fit of furniture and good design practice, e.g. don't put the cooker in the corner where you can't get at it), a library of reusable design ideas (expert kitchen designs) and a critic that gives suggestions when it detects the user has made a poor design (i.e. infringes a rule such as putting a cooker next to the fridge). Requirements are discovered by the active process of design. But requirements in this case are synonymous with the design; furthermore, one can argue that the real requirements are in the embedded rules. DODEs blur the distinction between requirements and design. However, this approach may be appropriate in a stable domain in which design is about versions that improve on an existing product, for example engineering domains such as vehicle, aircraft or ship manufacture. Designers in these industries build complex (interactive) simulations of products in development environments with iterative improvement of prototypes. While domain-oriented development environments are the exception rather than the rule, they are an interactive representation that goes far beyond our initial starting point of simple prototyping tools. In doing so, however, the point is made that interaction helps discovery-based learning and requirements analysis in some domains and is closely linked to refining designs.

After this digression, a summary of the essential cognitive properties of passive, active and interaction representations follows:

- *Passive representations*: examples are lists, diagrams, and all paper-based specifications. These have to be actively read or inspected. Their value depends on how well the form helps a particular task; for example, can a pathway be easily traced on a diagram. The disadvantage of passive representations is that they can encourage sloppy reasoning; people might agree with a specification without really thinking about it. The advantage is that they are cheap to produce.

- *Active simulations*: these representations can be played or run in some sense. The simulation follows a script and exhibits behaviour. The advantage of simulations is that they engage human reasoning more effectively, because we are challenged to interpret what we see. Also, several variations can be run to support comparison; however, the user experience is still passive as there is no need to interact.

- *Interactive prototypes*: these are active but also interactive, so users can test them by running different commands or functions. The degree of interactivity depends on the implementation cost. At the low end a set of scripts are provided that simulate a small number of commands, and departure from the pre-set sequence is not allowed. This variant is often called concept demonstrators. More expensive are narrow, in-depth prototypes in which functionality is fully implemented but only in a restricted sub-system, or broader, shallower prototypes that have functionality partially implemented throughout the system. Cost increases as prototypes converge with a fully functional product. Active engagement is guaranteed because the user has to operate the prototype and interpret what happens.

So far it may seem that interactive representations that are also comprehensible, and easy to learn and change, are bound to be more effective; however, there are other constraints to be considered. In the next section, more basic properties of representations are examined: the media on which they are displayed, and the physical artefacts that contain them.

5.2 Representations and Information Requirements

Since much information is shared among RE tasks, it is appropriate to analyze information requirements as a whole rather than partition them by task. First we start by considering the types of information that have to be gathered during requirements analysis.

Requirements are only part of the information that has to be captured during requirements analysis. Facts have to be acquired for modelling, scenarios and understanding the system context. Information can be classified into ontologies or logical categories, e.g. causal, temporal, sequential (Arens, Hovy and Vossers, 1993); however, for RE a simple classification will suffice. At the top level, information relates to either static (non-changing) or dynamic (changing) phenomena. Information may also describe physical, concrete phenomena that are perceivable parts of the real world, or abstractions that are created in software engineering to describe more generalized views of natural phenomena in models. Both of these top-level distinctions are closer to dimensions than categories; for instance, scenarios which are supposed to be descriptions of concrete real worlds may contain abstractions of agents and their roles. The types of information that need to be captured during requirements analysis can be summarized as follows:

- *Dynamic information*: events, actions, procedures and tasks. Tasks and procedures are composed of actions organized in an order by control constructs such as sequence, selection, iterations, concurrency, etc. Three sub-classes of dynamic information are distinguished: process, event life history and interaction. Tasks describe high-level activities, while event life histories describe action at lower levels of granularity, and interaction describes behaviour between two or more agents. User-system interaction falls into this category.
- *Static information*: entities, objects, agents, attributes, relationships, properties and states. Objects and agents are hybrid types as they contain both static (attributes) and dynamic (procedures, methods) information.
- *Contextual information*: This describes the system environment and contains both static and dynamic components:
  - *Scenarios*: there are many definitions of scenarios, but most assume examples and specific descriptions of behaviour and facts accompanied by the setting in which behaviour takes place.
  - *Scripts* are sub-sets of scenarios that give the events that happened without elaborating the context.
  - *Episodes* elaborate the context of a scenario with details of the physical setting in which an event or action takes place.
- *Intentions*: goals and intentions exist as plans and objectives held by individuals or organizations. The arguments and justification for the goals have been included in this group.

RE needs to gather information in all these categories (see Table 5.1). Goals give user requirements, while both static and dynamic information refine requirements and provide the subject matter for modelling. Contextual information may contribute to modelling and requirements but does so indirectly by giving the background for interpreting how the system should be designed. All categories contribute to models of the intended system, represented in familiar data flow, entity relationship diagrams, and object-oriented notations.

Domain knowledge belongs to the system environment but also includes examples, scripts and scenarios that describe instances of behaviour and physical details of objects. Episodes are composed of events with information about the context within which the state change took place – for example, if the event were an accident, the episode contains details of the scene, time, and people agents involved in the accident. The fourth category holds information that is generated by and

Table 5.1 Information types, properties and mappings to representations.

Information category	Contents	Properties	Typical RE representations
Dynamic information	Events, actions, procedures, tasks	Abstract models	Entity life histories, state transition diagrams, structured texts
Static information	States, attributes, entities, relationships	Abstract models	Entity relationship diagrams, lists
Context, environment, domain	Episodes, scripts, scenarios,	Physical examples	Scenario narratives
Intentions, goals	Arguments, decisions, goals, problems	Language-based facts, models	Goal trees, design rationales, requirements lists

belongs to stakeholders in the process, i.e. goals that describe users' wishes for the new system, problems with the current system and arguments about design. Argumentation relates to a meta-discourse about requirements and initial designs and contains propositions, supporting evidence, justifications and decisions.

Information properties describe the level of abstraction of the information content. Physical information includes visual or spatial properties of the world, such as structures, visual attributes of objects, pathways, spatial distribution, location, size and shape. Context information in domain knowledge is usually physical. Abstract information does not have a physical existence, for example knowledge, concepts, plans, propositions, values, and requirements stated as goals and models.

Information can be classified in more detail by using types as illustrated in the decision tree in Figure 5.2.

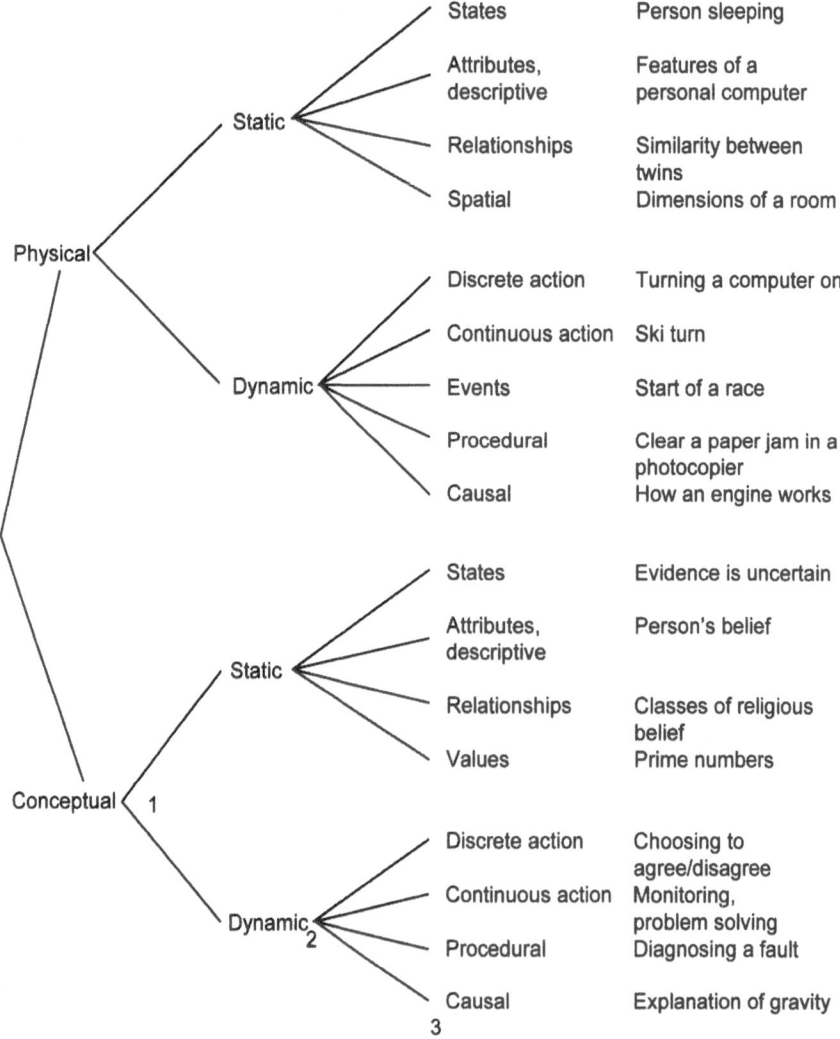

Fig. 5.2 Decision tree for classifying information types. The first decision point reflects abstraction from the real world, the second points to change in time and the third categorizes content.

This classification can prove useful in analyzing information requirements in their own right. Information has been much neglected in RE; however, in many functional requirements information constitutes the users' needs. Considering what type of information the users need for their tasks or for decision support is important, and the types point to primitives that will be represented in models (actions, events, states, attributes, objects) as well as domain knowledge and explanations (spatial, causal). The distinction between conceptual and realistic points to specification modelling (conceptual) versus scenario- and example-based realistic knowledge. Media choices then follow from these classifications.

5.3 Media and Representation

So far we have described the necessary information that RE representations should contain, and the demands that RE tasks and dialogues make upon those representations. The next issue to consider is choice of the media that convey a representation. RE has tended to be conservative in its choice of media. In most cases text and diagrams have sufficed; however, video has been used for recording meetings (Carroll *et al.*, 1994b) and video scenes have been captured for analyzing user tasks (Haumer, Pohl and Weidenhaupt, 1998). This section explores the range and combination of different media that may be employed for representing requirements. A taxonomy of media types, based on the forthcoming ISO standard (Sutcliffe and Faraday, 1994; Sutcliffe, 1999c), is used to explore the representational possibilities for an information need (Figure 5.3). Note that these definitions are non-orthogonal, so speech is both an audio and a linguistic medium.

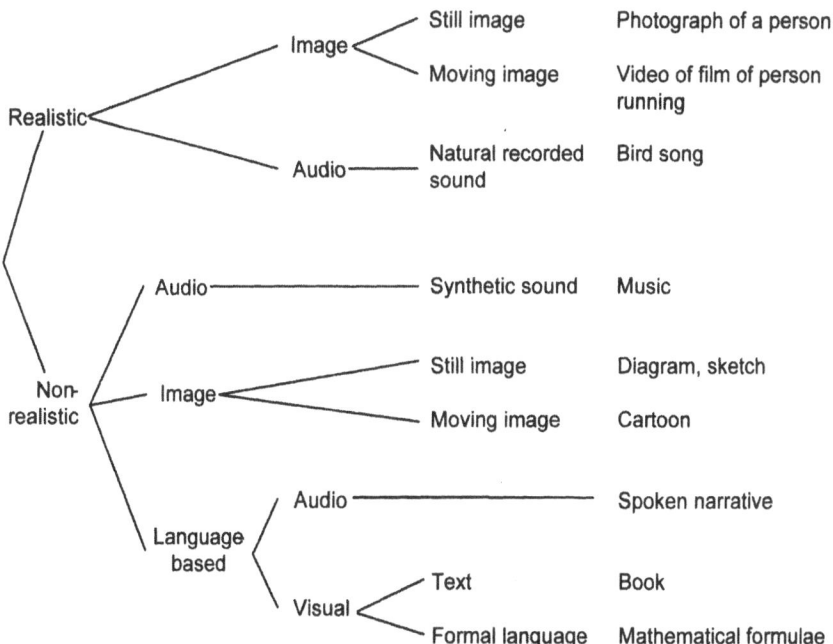

Fig. 5.3 Decision tree for classifying media resources.

Media types refer to the form of representation as it is perceived by the user. Definitions of multimedia abound, but for the purposes of RE we can take a simplified view of common sense distinctions.

The first point is that different media help the representation of *concrete or abstract* facts. Concrete facts are descriptions of the world (a photograph of myself); abstractions are produced by analysis, modelling and design (class definition of "Professors", of which I am an instance). Realistic representations are captured directly from the world; abstract or designed representations have to be created by human action.

The second distinction is between *linguistic and non-linguistic* media. Linguistic media, text and speech, represent language. Non-natural symbolic languages of mathematics and domain specific languages (e.g. circuit design in electronic engineering) also belong to this category. Non-linguistic media is everything else: images and sounds from the natural world. Diagrams pose an interesting boundary problem; in one sense they are symbolic languages, but informal sketches have few conventions and most people would say they are (semi-) natural representations of the world.

The final distinction is between *static and dynamic* media. Static media persist over time; they are inspectable and viewable, such as text or still images. Dynamic media, in contrast, are played continuously (e.g. moving image, sound and speech).

These distinctions are important because of two psychological properties. Dynamic media attract our attention and are difficult to ignore (try to avoid listening to the radio and concentrate on a newspaper instead). By virtue of their attention-grabbing powers, dynamic media engage our reasoning. The disadvantage is that dynamic media continually overwrite working memory because it has a limited capacity. The reason why you can't remember more that 10 per cent of an interview is that you have to make sense of one sentence, and store your understanding before the next sentence floods in. For the same reason we only remember the gist (top-level story line) and a few vivid scenes in a film. So we need static media to act as an external memory during problem solving. These distinctions lead to the media selections shown in Table 5.2; for more detailed classifications of media and complete methods for selecting and integrating multimedia representations see Sutcliffe (1999c) and ISO 14915 (ISO, 1998).

Some media choices are constrained by the properties of information, for example requirements lists and scenario narratives have to be conveyed by natural language text or speech. Abstract information may be subdivided according to its mode of expression, either informal using natural language, sketches, images and diagrams or formal expression with a set of mathematically defined semantics. However, there is a considerable space of media-representation type choices, as illustrated in Table 5.2.

Diagrams are probably the most popular RE representation, but their psychological properties point to some important limitations. Diagrams are designed images with language captions that have to be translated into a model. For instance in data flow diagrams (DFDs) the facts recorded in this notation can be interpreted in several ways. The data flow does not describe the order of arrival of the messages, or the preference in processing order. This would require queues and schedulers to be specified. In Figure 5.4, the DFD does not make it clear how loans and reservations are processed, whereas the entity life history shows a specific step for checking for reserved books before granting a loan. The semantics of DFDs can be improved by

Table 5.2 Properties and limitations of representational media. For more detailed classifications of media and complete methods for selecting and integrating multimedia representations see Sutcliffe (1999b) and ISO 14915 (ISO, 1998).

Medium	RE example	Created by	Limitations
Text: narrative	Scenario description	Written by interviewer/user	Potential ambiguity, slow processing
Text: lists	Requirements list	Recorded by interviewer	Only expresses ordering or group membership
Text: formatted	Tables, indented procedures	Designed by analyst	Understanding may be limited by format
Alphanumeric text	Values in specifications	Recorded/calculated by analyst	Denotation of values needs explanation
Still image: realistic	Photographs, scanned images	Captured from real world by cameras	Can't show behaviour effectively
Still image: designed	Drawings, diagrams, sketches	Drawn by analyst	Denotation needs to be explained
Speech	Voice notes, meeting recordings	Recorded on audio tape	Ambiguity, incomplete utterances
Sound	Domain noises, alarms	Recorded on audio tape	Depends on quality of recording
Moving image: realistic	Video of work domain	Video recording or film	Interpretation limited by scope of shot
Moving image: designed	Animated specification	Simulation designed	Interpretation depends on scope of animation
Interactive media	Prototype	Designed	Cost of implementation hides details

adding coding for event and control flows and symbols for processing order, but these are sticky-tape solutions on an ambiguous notation.

The two key design issues are defining the information content for a particular task, and accommodating non-functional requirements. The solution to this dilemma is to provide a concise notation that is not misunderstood and is amenable to logical reasoning, with an informal diagram to ensure that the representation is easy to comprehend. For example, a formal language such as Z (Spivey, 1989) may be used to express state transition semantics; alternatively, the same semantics can be represented, albeit less precisely, by a Jackson diagram. Validation tasks that require communication with users bias choice towards informal diagrams, whereas formal notations are more suitable for verification tasks that involve establishing truth properties of a model. Formal languages can be used for validation as well as verification; however, the formal representation has to be either explained in detail to the user or converted into simulations so that the consequences of the specification can be assessed (see Albert simulation tool, Dubois *et al.*, 1997).

Most RE representations are conservative in using natural language texts and diagrams but this limitation need not apply in the future. For instance, audio is economic to capture, so it makes a good medium for voice notes or annotations and it provides broadcast speech; however, its disadvantage is non-persistence, so it is unsuitable for conveying facts that need to be discussed and compared. Moving

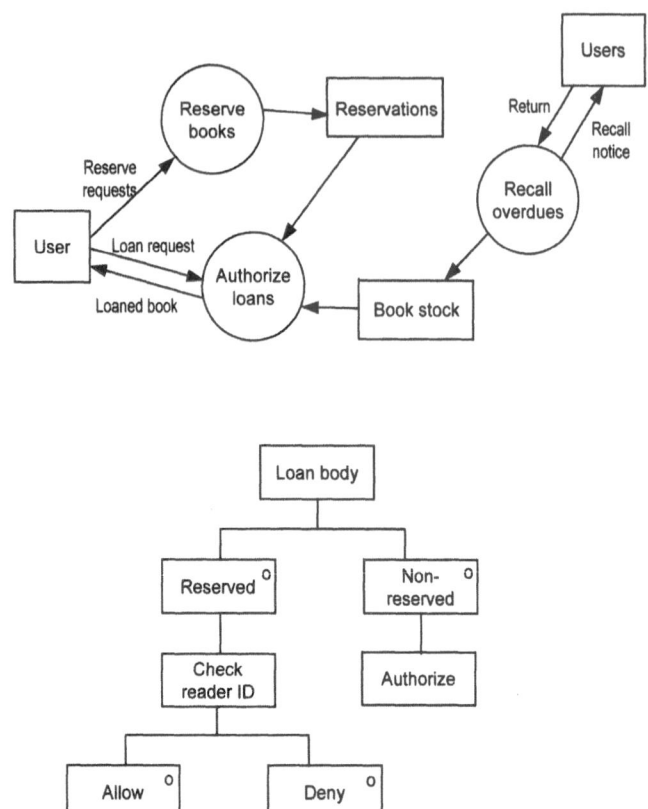

Fig. 5.4 Data flow diagram to illustrate the ambiguity of flow, followed by a more precise entity life history diagram.

image media are useful for conveying visual information on contextual and complex procedures. Video may be useful for capturing context aspects of domain
knowledge and recording decisions in design meetings (Carroll *et al.*, 1994b) but it
cannot communicate detail easily since video must be paused for inspection. An
important point to bear in mind with combining media is the difference between
static and dynamic media. Static media persist. People can refer to text and still
images throughout a session; hence these media act as aide-memoires. Dynamic
media attract attention and hence encourage engagement, but once they have been
played our memory rapidly fades. So we need dynamic media to help our reasoning,
with realistic media to stimulate our reaction by reference to the real world, and
static media to remind us about the key points and to record conclusions.

Static representations can convey a wide variety of information as specifications,
models, lists and narratives, but the user has to actively extract information. Active
presentations, on the other hand, engage the reader. Moving images express behaviour effectively but interactive media, for example prototypes and interactive simulations, help us to understand the implications of requirements by testing and
problem solving. Further guidance about the affordances of media types as suitable
representations can be found in Sutcliffe (1999b).

RE has relied on lists, diagrams and formatted text specifications, but video has been used (Carroll *et al.*, 1994b; Haumer *et al.*, 1998) to record meetings and capture records of real world behaviour. The limitations of narrative text are well known, as requirements written (or spoken) in natural language are inherently ambiguous and prone to misinterpretation. On the other hand, specifications in formatted text can be precise but pay the penalty in requiring the user to learn a convention to understand the semantics of the representation. The following section investigates the merits of different representations and media support for different RE phases and tasks.

5.4 Choosing Representations for RE Tasks

This part of the framework is based on the task model of RE described in Chapter 3. From the viewpoint of communication, RE involves gathering and clarifying information; presenting and explaining models; critiquing and comparing specifications and designs; and demonstrating and interacting with models and prototypes.

5.4.1 Requirements Acquisition and Fact Capture

This task acquires descriptions of system objects, components, functions, procedures, events, actions, domain facts, users' goals and requirements. Traditional support consists of lists, informal notes and diagrams; while fact-gathering interviews may be tape recorded to provide a more comprehensive report. However, image media can prove useful for recording behaviour and context in the domain. Video and photographs can supplement observations of users' work practice and the system environment. Audio recordings are useful records of informal communication between people and for levels of background noise that may affect requirements. The merit of capturing data in realistic media is that the wealth of physical detail can often indicate problems or supply facts that are important for subsequent analysis. Image and sound recordings can remind users of what they actually do when proposing a new design in requirements validation.

The disadvantage of video recording is volume. Even if resources are available to shoot sufficient video to comprehensively cover user behaviour, the analysis time would be prohibitive, even by simple eye-balling of the recordings. Accordingly most authors have proposed selective capture either of real world scenes that represent key sequences of behaviour (Haumer *et al.*, 1998) or critical segments in meetings when decisions are made, possibly linked to footage of context (Carroll *et al.*, 1994b; McDaniel, Olson and Magee, 1996). However, this still leaves the requirements engineer with the problem of selecting the sequence to record in the first place. Probably the best advice is to shoot video selectively to capture typical work procedures and the workplace, and then selectively record user behaviour that is linked to problems or areas of work that require detailed investigation. Video helps to share a feeling of the system's usage with others in the development teams. It can also be useful in presentations to set the scene for a new system proposal. As well as video and audio, fact capture also needs lists and narratives to record data.

5.4.2 Analysis and Modelling

Information requirements in the early stages of analysis are for high-level domain details, ideas to create the new system vision, with high-level user goals that may be derived from mission statements, policies, aims and objectives in the business. Scoping is generally supported by lists, informal notes and diagrams, as ideas and facts have to be easily comprehended by users. Informal sketches and diagrams may also be used to debate the system scope, for instance by moving the system boundary.

Requirements analysis has borrowed representations from structured system development methods, such as data flow diagrams, entity relationship diagrams, lists and hypertext, as well as formal models imported from software engineering (Loucopoulos and Karakostas, 1995; Jackson, 1995). Informal modelling notations, such as use cases and UML diagrams, have been widely used, with hypertext for informal structuring of linguistically expressed requirements (Pohl, 1996). Rich picture diagrams (Checkland and Scholes, 1990) may also be used for informal modelling. These map out the relationships between people in the system (agents), the work or tasks they carry out, and the locations or organizations where they work. Since there is no specific notation for rich pictures they can be misinterpreted, although the informality helps communication with users. An alternative approach is to use prototypes or concept demonstrators that not only record the model but also allow the user to interact with it. Analysis and modelling are frequently interleaved to elaborate the requirements as understanding of the problem domain increases through the act of representation. Linguistic media play an important role in analysis and modelling by recording facts and requirements as lists, tables and in narrative form as texts and annotations on diagrams. Establishing relationships is a key need that involves recording, editing and presenting a representation to others. Representations should not only facilitate easy recording of information but also allow incremental construction and editing.

Modelling transforms concrete facts into abstract interpretations, and helps to summarize collections of facts and their interrelationships. Although these can also be represented in formatted text, diagrams afford easier visual scanning and tracing of pathways. Diagrams are good for summarization as models of abstract facts but there is a trade-off between the precision of semantics and comprehensibility. Context diagrams (De Marco, 1978), use case diagrams (Jacobson *et al.*, 1992) or rich pictures (Checkland, 1981) all provide informal representations at an early stage to help system scoping. Informal diagrams are popular because they are simple, but carry the penalty of hiding considerable ambiguity in the notation, as illustrated in Figure 5.5, adapted from Beyer and Holtzblatt's (1988) contextual design method. Some might call this a physical model diagram, and others a drawing. More complex diagrams, such as semantic data models, require considerable learning to interpret but repay the reader with more precise meaning. More precise notations, such as those used in semantic data models (Verheijn and Van Bekkum, 1982), have proved less popular, so a better combination may be to use informal notations for ease of communication coupled with tools for formal representations that enable automated reasoning.

Accuracy becomes a more pressing concern in modelling, so ideally a representation should convey a precise meaning. More formal specifications use symbolic languages and formatting conventions but pay a price in terms of accessibility to users; hence more complex representations can only be shared between developers.

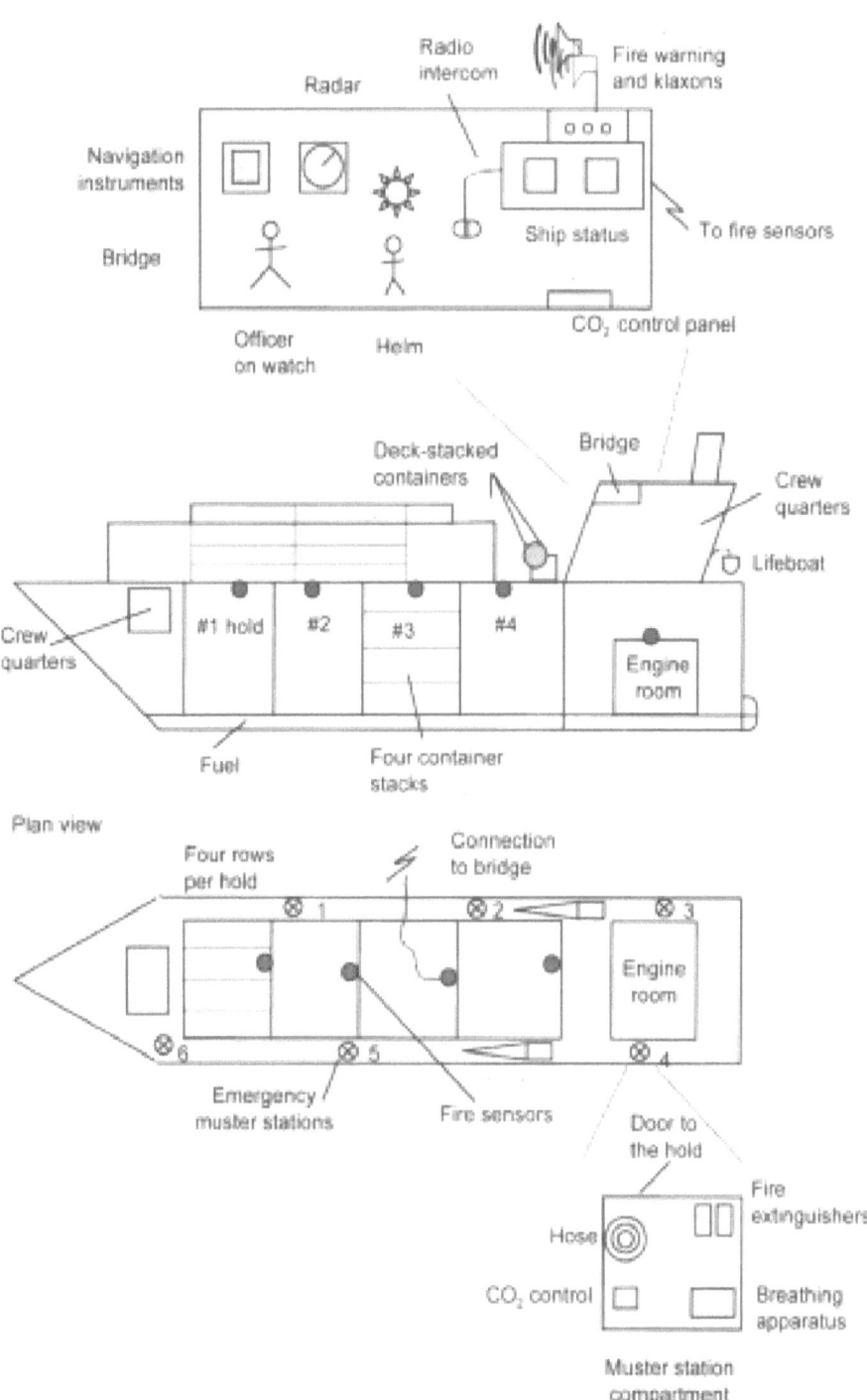

Fig. 5.5 Domain model diagram for a ship-board emergency management system.

5.4.3 Validation

This task requires models of the desired system expressed either in a notation or as prototypes; however, it also needs information about the system context. Domain knowledge may be held by users, who can check the requirements; alternatively, domain facts may be explicitly represented for validation. Walkthroughs are a common approach in which semi-formal specifications such as data flow diagrams are critiqued and discussed in a workshop of designers and users.

In validation, representations need to establish a common lingua franca for shared understanding between users and developers. This restricts the use of formal specifications to experts who are familiar with the notation, but care has to be exercised with informal diagrams because users may think they understand a notation when in fact they don't. Validation needs to actively engage users in checking that the designed system actually does what they want, so active and interactive representations are important. Simulations and animations engage user attention using scenario-based representations to help users see the implications of system behaviour and thereby improve validation (Johnson *et al.*, 1992). Speech and sound also encourage reaction, while the analyst can bring static diagrams to life by a walkthrough, pointing out what happens to dependencies between objects, tasks and agents, etc. This can be effective, but simulation and interactive prototypes make validation easier.

Paradoxically, RE started with concrete data and progressed to abstractions; in validation we need to return to the concrete. People have a tendency to misinterpret or ignore abstract representations. Abstractions are understood by those who create them because they went through the mental effort of induction (from the specific to the general). Unfortunately many representations, especially informal ones, allow people to put their own concrete interpretation on a generalized abstraction. Misconceptions about what is allowed or how a new system will work can creep in as individuals take their own (often unverbalized) impression away from inspecting a passive representation. Formal specification is no panacea, as most users will never expend the effort of learning another foreign language – for that is what a formal specification is.

The solution is to mix the concrete and the abstract. Concrete representations promote reasoning by anchoring arguments in people's memory and experience. When challenged with the familiar we react. Everyone will have an opinion or an example, story or episode. In contrast most of us are happy to take the opinion of experts on the generalities of policies, objectives, roles, etc. Only when the implications become apparent in real practice do we object. Consequently, representations of concrete facts, behaviour and usage context by video, photographs, speech and text narratives help in requirements validation. But this is not just representing the past; concrete media may also be used to show concrete visions of the future. Realistic media are a key component of the success of scenarios (see Section 5.6). Scenarios describe the context of use for a system as well as providing a vision of how the system will work. People tend to react to specific examples presented in scenarios, but this presents another dilemma. A single scenario shows only one possible sequence of events among many possible sequences permitted in the abstract specification. Hence we need a set of scenarios represented as animated sequences of behaviour that provide good coverage of the specification, and the analyst needs to cross reference the concrete to the abstract, for example "At this point we have shown you options X and Y, but you could also complete the task by options W or Z".

An improvement on animations is to use interactive representations: prototypes or concept demonstrators that evade (although not completely avoid) the problem of multiple scenarios by allowing the users to explore alternative paths.

5.4.4 Negotiation

The essence of this task lies in the discussion, explanation and handling of conflicting requirements. Stakeholders often hold different requirements, so negotiation involves comparing, prioritizing and deciding between the different requirements or design options. Ranked lists or decision tables are useful for this task. Functions, design options or goals form the information input to this task, with criteria for ranking or prioritization supplied by the stakeholders. Context and domain information may be necessary for interpreting and resolving conflicts. Models and lists portraying different views are necessary, with multiple copies to represent different stakeholders' viewpoints. Representations that facilitate comparison between different viewpoints, i.e. tables and matrices, are useful.

The key properties of media combination for each requirements task are summarized in Table 5.3.

So far we have investigated media and the cognitive desiderata of representations, but that does not describe the display devices on which representations are delivered. The properties of devices can radically alter their effectiveness for different RE tasks; for instance, a paper-based list can be shared and passed about a meeting, whereas a computer-displayed list has to be projected. Broadcasting and sharing information is necessary for collaboration in RE but it can be expensive and sometimes less convenient. In the following section we turn to these issues.

Table 5.3 Representations and their affordances for RE tasks.

Requirements task	Media types	Properties	Justification
Acquisition and fact gathering	Video, speech, sound, photographs, text lists	Concrete representations	Capture of detail
Analysis and modelling	Formatted texts, diagrams, lists, tables	Abstract, static representations to support induction	Linguistic conceptual expression of models and designs
Validation	Diagrams, lists, formatted texts combined with video, images and text	Mix of concrete and abstract, also dynamic and static representations	Engagement for problem solving with aide-memoires and grounding to anchor reasoning
Negotiation	Diagrams, lists, formatted texts, decision tables	Abstract representations of choices, trade-offs, criteria and arguments	Format to structure argument and options, linguistic expression plus value for ranking

5.5 Delivering Representations on Artefacts

RE implies CSCW (computer supported co-operative work) because it is a collaborative activity, so we cannot avoid the issue of how representations function in groups. Technology changes rapidly, so it becomes harder to describe properties that will hold true as technology advances. For instance, flat screen portable displays and data projectors have improved the ability of computers to deliver different media in a broadcast manner. Consequently the approach is to draw out a set of criteria by which the affordances of delivery technologies can be judged, and then take a snapshot of currently used devices for comparison against the criteria.

The overall effectiveness of a representation depends on the information represented in one (or more) media rendered by a device. The following affordances affect the suitability of a technology in collaborative RE tasks:

- *Visibility:* some devices, such as most computer screens and pieces of paper, have small display areas, whereas overhead projectors and whiteboards have a larger visible area. A larger display area allows more information to be communicated, or information can be represented more clearly by larger type or magnifying images. Visibility interacts with shareability.

- *Shareability*: this is a crucial property in collaborative interaction. Shareability has two senses. First, a device may make information shareable in the broadcast sense that all participants can easily acquire the same information – for example, large screen projection. The second sense is more complex. A representation may be shareable because all participants can refer directly to components in the representation by pointing. This "deictic" shareability is a function of proximity to the representation and may be limited by pointing devices (e.g. physical laser pointers).

- *Portability*: the size and mass of a device dictates how easily information may be transported between people or meetings; for example, a diagram on a piece of paper is very portable and shareable, whereas a flipchart sheet is portable, although not conveniently so. Information portability in computer tools depends on the system design, such as single user versus networked, and the storage media, for example diskette versus the Internet.

- *Retrieval and manipulation*: devices allow information to be represented, stored and retrieved. Since computer systems are functionally rich, they provide large-scale storage facilities as well as retrieval mechanisms. In contrast, flipcharts offer few affordances beyond limited storage. Representations may be easy to change or require some effort to modify according to their host device. Altering a diagram on a whiteboard is not too difficult as it can be erased, but on a flipchart a new version must be drawn.

- *Representation affordances:* computer systems have good representational affordance for multimedia, although other devices such as 35 mm slides and video can represent text, diagrams, and moving image. However, the functional affordance of capturing and representing behaviour on film is not as effective as interactive computer-based simulation.

With this checklist, commonly used devices can be evaluated for the constraints they place on the communication goals in RE tasks. Visibility and shareability are important in most tasks where participants in the conversation need shared access to diagrams and lists conveying representations of requirements, models and rationales.

Whiteboards and flipcharts are highly visible and facilitate information sharing; however, their representational affordance is low. This usually limits their use to representing lists of requirements and simple models. Paper can be used for representing user goals, informal and formal models, and can capture some domain knowledge as text, diagrams and other images, but it is less visible to all participants, although it can be shared. Computer prototypes may not be very visible, unless the screen is projected, and this has an impact on their shareability. Prototypes are reasonably portable on laptops, and web technology is reducing that limitation; furthermore, their main advantage is in rich functional affordance. This allows models of the required system and its design realizations to be represented as scripted simulations and interactive prototypes.

5.6 Representational Paradigms

Having compared representations at the level of media and delivery devices, the final step is to complete the picture by considering more complex or compound representations that are in current RE practice or have been proposed in research. These representations can be termed "paradigms" as they are frequently composed of several different media or individual representations. The paradigms considered here are:

- *Informal models and records*: these are composed of media that are realistic or abstract but cannot portray precise information. The test for informal models is that most people can understand them with minimal or no training.
- *Formal models*: abstract representations that require considerable training to be understood. Formal requirements specifications (e.g. Van Lamsweerde *et al.*, 1995) fall into this paradigm.
- *Scenarios*: this is a heterogeneous paradigm related in its various interpretations to both informal and formal models, although the essence of scenarios is representing concrete facts. It is treated here as a distinct paradigm because of the interest scenarios have attracted (Carroll, 1995).
- *Prototypes*: simulations, concept demonstrators and other interactive models are included in this category.
- *Design rationale and claims*: these could be regarded as informal models, but unlike other paradigms design rationale and its relatives record design decisions or design knowledge rather than requirements *per se*.

Clearly technology and software implementation affect the nature of each paradigm. Software tools enable easier access to any representation and can afford more interactivity. Accordingly, the current level of tool support for each paradigm and prospects for future improvement will be considered.

5.6.1 Informal Models and Records

Informal models and simple recording of requirements use text media as lists and narratives combined with informal diagrams and sketches. Storyboards are a well-known means of informal modelling in multimedia design (Fisher, 1994). Originating from animation and cartoon design, storyboards are a set of still images that represent key steps in a design. Translated to software, storyboards depict key stages in interaction of a product. The next step is making storyboards more interactive by conducting walkthroughs to explain what happens at each stage, but this is still a passive representation. Allowing the users to edit storyboards and giving them a construction kit to build their own encourages active engagement and RE by doing. This is the essence of the PICTIVE method (Muller, 1991). Other researchers are trying to extend the flexibility of storyboards towards "paper prototypes" (Wilson *et al.*, 1997) which give a construction kit so that rough and ready sketches can be scanned into a computer and then linked in a script to form a limited animation. This approach converges with building concept demonstrators using multimedia or hypermedia authoring tools (e.g. Macromedia Director) to rapidly develop semi-interactive prototypes. Most RE tools support this paradigm by managing lists for domain facts (objects, agents, actions, etc.) and requirements, for example DOORs, RequisitePro and Cradle. However, support for multimedia and animation is poor even though these representations are necessary for informal models. There is considerable potential for developing RE tools that integrate a requirements database with multimedia documentation support and development tools for storyboards/concept demonstrators. This could offer seamless support for building requirements demonstrators, evaluating them, and recording the requirements arising, until complete prototypes are developed.

5.6.2 Formal Models

Just what constitutes a formal model is a matter of some debate. Industrial practitioners often regard diagrammatic notation as formal, for example the event process control diagrams of SAP ERPs (enterprise resource plans) or activity sequence diagrams from UML. Academics from computer science would disagree. Their view of formality implies a rigorous mathematical basis underpinning the notation of the specification language. Examples are state-based languages such as Z (Spivey, 1989), logic-based languages in various forms (e.g. modal action logic) and process algebras (e.g. LOTOS). Most formal models use formatted text to express specifications, possibly combined with a less formal diagrammatic representation. Less formal members of this paradigm highlight the diagram representation first and the structured text second (e.g. entity life histories in JSD (Jackson, 1983).

Formal models are passive representations that are difficult to understand. However, they can benefit greatly from computerized animation to show behaviour pathways triggered by specific events, agents involved with actions, etc. An example of such a tool is Albert II (Dubois *et al.*, 1997) which animates event sequences in an agent model based on a formal specification of temporal semantics. Similarly the GRAIL tool supports interactive querying and simulations of specifications written in the KAOS language (Van Lamsweerde and Letier, 2000). Requirements specifications in formal models are easy to implement because their semantics are precise. This brings added benefits of automated checking and verification to ensure that a specification is consistent and correct. Consistency is checked by searching the model to make sure specification rules are not offended – for example, every goal must have a task that achieves it, every task must have an agent responsible for it.

Correctness is more difficult. Without going into a treatise on formal methods, theorem provers can be run to check properties such as invariants (the systems can never get into a specific, usually dangerous, state); deadlock (two processes hang because each is waiting for input from the other); liveness (when specific conditions apply a process will always be active); and reachability (a state can be activated by following at least one pathway through a specification) (Milner, 1989).

5.6.3 Scenarios

Unfortunately the term "scenario" has been hijacked by too many authors to have any commonly accepted meaning. The *Oxford English Dictionary* defines a scenario as "the outline or script of a film, with details of scenes or an imagined sequence of future events". In RE a scenario can have three quite different definitions:

1. A story or example of events taken from real world experience. These stories are close to the common sense use of the word and may include details of the system context (scenes).
2. A single thread or pathway through a model (usually a use case). This is the sense in which the object-oriented community use the word (Cockburn, 2001; Jacobson *et al.*, 1992; Rational Corporation, 1999).
3. A future vision of a designed system with sequences of behaviour and possibly contextual description. In this case the scenario comes close to a design mock up.

Depending on one's usage, scenarios are represented in different media. Type 1 scenarios are captured as speech or text narratives and may be embellished with still or moving images to illustrate the context. Type 2 scenarios are represented as formatted text and possibly illustrated pathways in use case or activity sequence diagrams. Type 3 scenarios can be represented by anything from a storyboard sketch to animated sequences in a concept demonstrator. The advantage of scenarios lies in the way they ground argument and reasoning in specific detail or examples, but the disadvantage is that being specific loses generality. The power of scenarios is their anchor in reality that forces us to address the "devil in the detail" during requirements validation. The problem is that we need many specific scenarios to test a general requirements specification. This can be summarized as the 20/20 foresight problem: how to capture (or generate) a sufficient set of scenarios to cover all the important problems in the application. Automated tools are needed to generate and then utilize large volumes of scenarios for requirements validation.

Tool support for scenario-based RE is still primitive. General tools can help to represent and store scenarios (e.g. word processors for narratives and graphics tools for sketches and diagrams). Type 2 and 3 scenarios can be represented in a variety of CASE tools for formal and semi-formal models as traces or simulated pathways through models. Support for generation is nearly non-existent apart from the CREWS-SAVRE tool (Sutcliffe *et al.*, 1998) that interactively generates Type 2 scenarios from use cases by applying a set of rules to suggest reasons why alternative pathways should exist.

5.6.4 Prototypes

Prototypes are interactive representations which, by virtue of embodying a computer system, use text and interactive graphics in the graphical user interface. However, prototypes are not a panacea for the RE problem and can waste resources when too many design ideas are explored without any agreement (Attwood *et al.*, 1995). Furthermore, prototypes hide the specification from the user and functionality may appear to be an illusion, so expectations are raised only to be frustrated in the final implementation. Breadth-first prototypes just give a look and feel overview, with a poor impression of system behaviour because little functionality can be implemented in any one sub-system. Depth-first prototypes provide the converse, while the zig-zag compromise between these two depends on spotting the key areas of the system where implementing more functionality will pay off in requirements validation.

Prototyping is well supported by tools ranging from programming languages such as Visual Basic and JavaScript, to multimedia tools that can be used to develop interactive but less fully functional concept demonstrators. Even though prototypes have the advantage of being an interactive medium that encourages user engagement, there is a sting in the tail for interactive representations. If users become too focused on the interaction they may comment on surface aspects of the user interface and fail to think about how the system would support their tasks.

5.6.5 Design Rationale and Claims

This paradigm is somewhat different because it represents decision and design knowledge rather than specifications or facts pertinent to the design itself. Design rationale is represented by formatted texts and simple diagrams that link issues (requirements or design goals) to positions (alternative solutions) and arguments (justifications for or against a position). Tool support comes as database managers with diagram editors (Conklin and Begeman, 1988; Buckingham Schum, 1996). The example in Figure 5.6 shows the design options for an issue in an information retrieval application with the argument expressed as criteria in the QOC (questions options criteria) variant of design rationale (MacLean *et al.*, 1991).

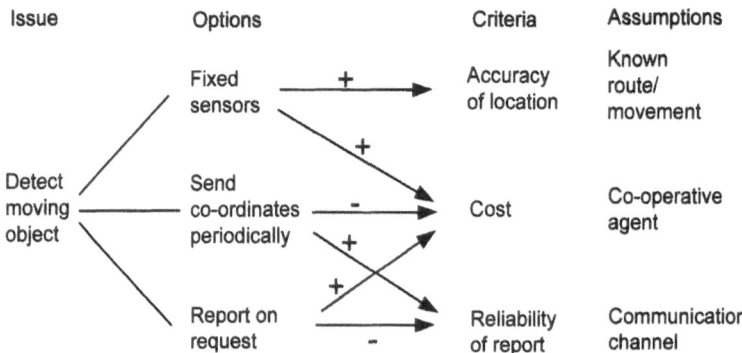

Fig. 5.6 Design rationale graph for a location tracking application.

Claims (Carroll and Rosson, 1992; Sutcliffe and Carroll, 1999) develop design rationale by structuring the criteria into advantages (upsides) and disadvantages (downsides). However, claims are a richer representation because they associate a requirements issue with arguments and a concrete scenario that describes the system's context of use. Further information is provided by an example product that illustrates an implementation of the requirements and design principles expressed in the claim. The distinctive feature of claims is the combination of abstract, model-based representation with concrete factual information to help interpretation of requirements. Claims only represent one position (a functional requirement) that is motivated by a design problem (the issue) described in a Type 1 (example-based) scenario. Arguments become upsides and downsides that express possible side effects of the claim. Claims also include an example or artefact that illustrates application of the design advice contained in the position, so this can be equated with a prototype. Informal or formal models may be linked to design advice and specifications as patterns to complete the integration provided by claims (Sutcliffe and Carroll, 1998); see Figure 5.7. However, there is as yet no tool support for claims.

Claim ID:	Preformed query claim
Author:	A.G. Sutcliffe
Artefact:	Preformed query library (INTUITIVE class) and UI menu selector

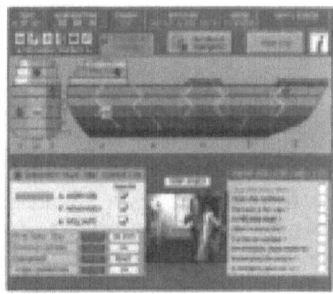

INTUITIVE shipboard emergency management system, showing the checkbox artefact (bottom right) and preformed queries as diamond icons leading to the answer box (bottom left).

Description:	a preformed query is attached to an icon or placed in a menu.
Upside:	provides rapid access to information; saves user's effort in articulating queries.
Downside:	finding queries in a large library can be difficult; queries inflexible, cannot be adapted.
Usage scenario:	the user interface presents a menu of queries which can be reused. The user selects a query that matches their need by selecting an icon or a menu option, and the system retrieves the information for presentation.
Effect	the user can retrieve information by reusing familiar queries.
Dependencies:	information needs and queries must be known, future information needs must be predictable.
Issues:	query format, customizability of values and constraint clauses, addressing queries to different databases.
Scope:	information seeking tasks with predictable needs.

Fig. 5.7 Example of a claim, showing the integration of concrete and abstract information to describe a requirement.

Table 5.4 Comparison of paradigms against the effectiveness criteria for representations.

Paradigm	Task affordance	Ease of learning	Ease of manipulation	Scope
Informal models	Acquisition, analysis	Easy	Easy	Breadth: good Depth: poor
Formal models	Modelling, analysis	Difficult	Difficult	Breadth: medium Depth: good
Scenarios	Acquisition, analysis, validation	Easy	Easy	Poor for any one scenario; depends on volume
Prototypes	Analysis, validation	Medium: depends on prototyping language	Medium: depends on prototyping language	Good, but depends on resource and testing
Design rationale and claims	Analysis, modelling, validation	Easy	Difficult: retrieval and change hard	Medium: depends on size of library

Claims assert trade-offs (upsides and downsides) that pertain to design features. Scenarios set claims in a specific context of use while generic models can be used to scope the applicability of claims in terms of application classes to which a claim may apply (Sutcliffe and Carroll, 1998).

Having reviewed the paradigms we can assess how each one fares against the desiderata outlined at the beginning of this chapter. Table 5.4 summarizes this assessment.

Informal models score well for ease of use and manipulation as they can be created and changed quickly; however, they suffer from poor representation of detail, even though a combination of different types of model can give broad coverage. More complex models which overlay several views, such as i* (Mylopoulos *et al.*, 1992), can give a broad coverage of goals, agents, tasks, resources and their dependencies; however, the diagrams become difficult to read. Formal models in contrast have the opposite profile: what they gain from in-depth coverage of detail they lose in difficulty of learning and manipulation, a consequence of the formal laws that govern such languages. Formal models are better at supporting verification and modelling tasks whereas informal models lend themselves to acquisition and analyzing requirements. In RE we frequently need both ambiguous and precise representations. In early analysis, requirements cannot be precise, and ambiguity needs to be tolerated to accommodate different viewpoints. To provide requirements freedom in expression (Fickas and Feather, 1995) we need informal models, but later in the process ambiguity and disagreements have to be resolved so that a design can proceed. At this stage formal models become more important. The key point is the transition between models. Either a range of representations might be used with a means of translating information between viewpoints (Finkelstein, Kramer and Nuseibeh, 1992), or animation tools can embed the formal representation and explain the consequence of system behaviour with less formal representations (Dubois *et al.*, 1997; Johnson *et al.*, 1992).

Scenarios provide many specific examples that can be generalized to create models during requirements acquisition, and thus also play a key role in validation. The advantage of scenarios lies in their concrete descriptions of the real world or visions of a future design that provide "reality checks" against which abstract models can be checked. Unfortunately the specificity of scenarios creates the problem of completeness. More research is required to discover new methods for generating or capturing a sufficient set of scenarios for requirements validation. In spite of these limitations scenarios are a useful representation for eliciting and communicating requirements. Claims and design rationale add design knowledge, which is not explicit in the other paradigms. Design rationale is useful for negotiation; however, the representation format is not as effective as simple decision tables, which can express exactly the same information. Claims augment the link between requirements and design knowledge by providing a context for its interpretation in the form of scenarios and prototypical designs. Claims, therefore, are the most sophisticated representational paradigm that can integrate other forms, from abstract models, to concrete contextual knowledge and design argumentation. The disadvantage of more complex representations may be an increased burden of learning; however, this will be more than offset by providing a means of integrating all the necessary knowledge for reasoning about requirements in analysis, validation and negotiation tasks.

5.7 Summary

The role of representations that support requirements engineering was investigated by first examining the information required by different RE tasks. This ranged from facts to requirements, models, domain knowledge, examples and scenarios. Information was then examined in terms of more abstract categories or information types, which are incidentally one way of categorizing information requirements. These types describe information as events, actions, objects and attributes, familiar semantic primitives in software engineering, but other types are more complex explanations for procedures and causation. Information types can be mapped to representations in different media following media selection guidelines drawn from the ISO 14915 standard. Requirements representations such as speech, image, text, lists, video, diagrams and so on, have different roles to play in tasks, according to the persistence of each medium and its cognitive properties. Representations were considered in a wider sense of the display technology that conveys a medium. A set of principles was described to assess the affordances of representational technologies: visibility, shareability, editability, portability and functional manipulations. Current computer and non-technological devices were compared with this framework to indicate where flipcharts, paper, whiteboards, etc. may contribute and computer displays might overcome the limitations of current representations. Finally, a set of RE representation paradigms were discussed in light of media selection and display devices. These were linked back to the information requirements of RE tasks, and included lists, models (both formal and informal), storyboards, prototypes and scenarios. The role of abstract and concrete representation in the reasoning process of RE was discussed.

6 Scenario-Based Requirements Engineering (SCRAM)

So far I have discussed the more theoretical aspects of RE, interspersed with some practical advice; in this chapter a practical method of scenario-based RE is described that incorporates many of the ideas expounded in previous chapters. The chapter also draws on experience in requirements evaluation with prototypes (Sutcliffe *et al.*, 2000; Sutcliffe, Economou and Markis, 1999). The method started life in research projects and has since been improved through several iterations to accumulate the lessons of practice. SCRAM has been validated by empirical study to test its strengths and weaknesses (Sutcliffe, 1995c, 1998; Sutcliffe and Ryan, 1998).

The method advocates a prototyping style of approach to RE, motivated by the need to get users actively engaged in reasoning about how a design will help to achieve their goals. Although SCRAM is applicable to a wide variety of systems it may not be suitable for large, complex and safety critical systems, where a specification-led approach is advisable. This is covered in the following chapter. SCRAM does not exclude formal specification and modelling of requirements, but it sees such activity as proceeding in parallel with prototyping.

6.1 Background

The approach is based on the hypothesis that technique integration provides the best avenue for improving RE and that active engagement of users in trying out designs is the best way to get effective feedback for requirements validation. Another motivation is to use scenarios as a means of situating discussion about the design, so that new requirements can be elicited by reasoning about problems posed by scenarios describing a context of use. Three techniques are used:

- *Storyboards, concept demonstrators* and *prototypes*: to provide a designed, interactive artefact that users can react to.
- *Scenarios*: the designed artefact is situated in a context of use, thereby helping users relate the design to their work/task context.
- *Design rationale*: the designers' reasoning is deliberately exposed to the user to encourage user participation in the decision process.

The techniques are combined with the SCRAM method to guide the requirements engineer. The method is composed of advice on setting up sessions, use of the above techniques, and more detailed guidance on fact acquisition and requirements validation strategies. The method consists of the following four phases:

- *Initial requirements capture and domain familiarization.* This is conducted by conventional interviewing and fact-finding techniques to gain sufficient information to develop a first concept demonstrator. In practice this takes 1–2 client visits.

● *Storyboarding and design visioning.* This phase creates early visions of the required system that are explained to users in storyboard walkthroughs to get feedback on feasibility.

● *Requirements exploration.* This uses concept demonstrators and early prototypes to present more detailed designs to users in scripted, semi-interactive demonstrations so the design can be critiqued and requirements validated.

● *Prototyping and requirements validation.* This phase develops more fully functional prototypes and continues defining requirements until a prototype is agreed to be acceptable by all the users. The prototype is then converted into the final product.

The method stages with inputs and outputs from each stage are summarized in Figure 6.1. It provides process guidance for conducting walkthroughs and organizing the requirements analysis process, with guidelines for interviewing and managing requirements conversations. Chapters 3 and 4 supplement this guidance.

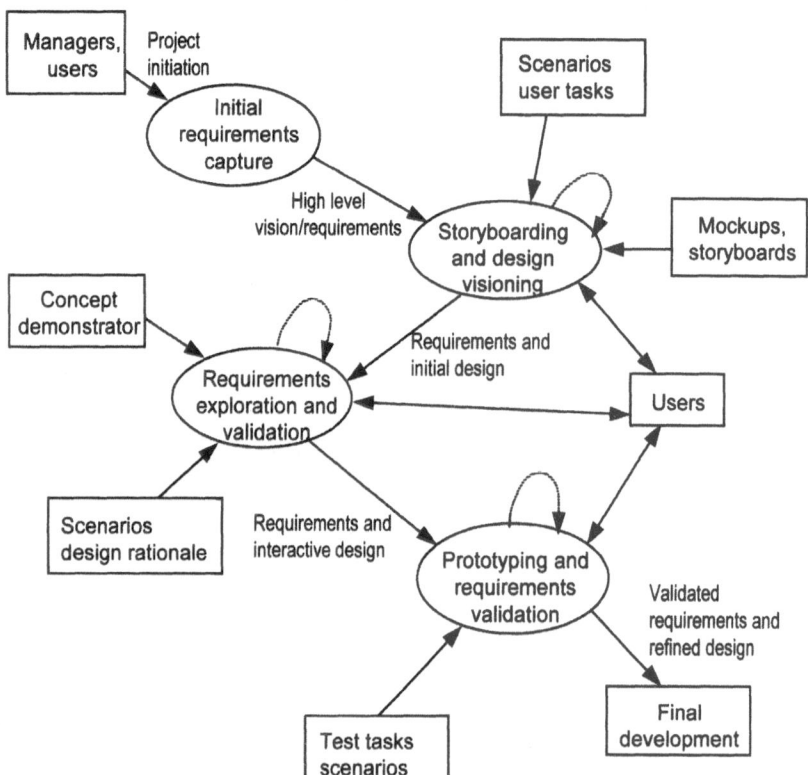

Fig. 6.1 Overview of the SCRAM method.

6.2 Initial Requirements Capture

This phase gathers facts about the domain and captures users' high-level goals for the new system. Interviewing and other fact-finding techniques are used (see Chapter 4) to scope the domain and acquire scenarios of current system use. Scenarios are elicited as examples of everyday use of the current system, with stories of problems encountered and how they are dealt with. Gathering a sufficient set of scenarios is a vexed question. There are several problems that might be encountered (see also Chapter 3):

- Users tend to miss out steps in scenarios that they assume are known to the analyst: the implicit or tacit knowledge problem.
- Each person may give an individual view of problems encountered. It can be difficult to distil a set of common problems from users with diverse views.
- Acquiring a sufficient set of scenarios to cover not only normal use but also situations when things go wrong can take considerable effort. The volume of scenarios can become daunting, and this presents a selection problem of finding a sub-set to use in requirements analysis.
- People tend to either forget abnormal examples or to exaggerate problems. Problems encountered most frequently and recently will be recalled first, but individuals will remember different episodes. Unravelling these potential biases can be difficult; for example, a personality clash may make a particular problem vivid for one individual, whereas for everyone else the problem was minor.

The best way to proceed is to gather scenarios of normal system use; look for commonalties between different individual versions and create a common "normal use case". Note that where individual variations occur, these can be useful hooks for questions later on about different individual strategies for using the system. Once the normal use case is in place, gather a set of exceptions (c.f. alternative paths in use cases). Note any problems motivating alternative paths, as these will often indicate requirements for the new system. The number of alternatives necessary depends on system complexity and safety criticality. Complex systems, when safety and security are important concerns, will require more scenarios than information processing systems. A sub-set of exception scenarios are selected for use in requirements analysis, but all the exception scenarios will prove to be useful in refining requirements even if they cannot all be used in interactive analysis.

This phase should also capture users' high-level intentions. Sometimes users initiate requirements analysis with a well-defined brief describing what they want. In other cases the investigation commences with a few high-level statements of needs. In this case brainstorming techniques and goal analysis are appropriate (see Chapter 4).

6.2.1 Goal Analysis

Scenarios complement goal modelling because, whereas goals focus on abstractions that describe users' intentions, scenarios make abstract intentions clearer by giving examples of how a new system might work to fulfil users' goals. The weakness of goal modelling is recording vague intentions without real thought about their practical implications. Two types of scenario can help by making the abstract, concrete:

- *System visions* express policies and high-level aims, and are frequently related to mission statements. While these are not scenarios in the sense of specific examples they do provide a framework within which users' intentions can be analyzed. To give an example, in safety critical systems the top-level policy might be "to expedite the safe navigation of the ship within the constraints of operational efficiency". Obviously this statement does not say what safety means nor does it explain the constraint of operational efficiency; however, the statement does provoke such questions as a starting point for the dialogue.

- *Impact scenarios* are visions of the future system's usage once the goal has been implemented in a design. These scenarios cannot be created until analysis has decomposed the system to a level where some detail of sub-goals is apparent. Impact scenarios describe the effects that the designed system will produce and hence test whether the goal is suitable. An example might be "the system diagnoses the potential fire hazards with different types of cargo and recommends safety measures to take if the cargo is likely to be explosive or emit noxious chemicals. The information is passed to the fire fighting crew who can take appropriate action." This scenario suggests further questions (and hence discovers further goals) about assumptions concerning the system's knowledge of the cargo – which may not be accurate – and whether the crew know what the appropriate action is (prompting a possible training requirement).

Scenarios therefore have their role to play in complementing goal models that record a hierarchy of user intentions and their relationships. As goal analysis often leads to changes in the analyst's view about the levels of a hierarchy, the ability to change hierarchical levels in a flexible manner is important.

The output from this phase is:

- a set of *user goals*, organized in a goal hierarchy if necessary;
- *domain facts*, captured as lists from interviews but also including sketches and possibly photographs of the system environment;
- a set of *scenarios* of use including normal and exceptional episodes, with a contextual description of the system.

These outputs are used to create early visions of the system in the next phase.

6.3 Storyboarding and Design Visioning

This phase creates early visions of the system as quickly and cheaply as possible. Storyboards are the most effective technique, although these can be augmented by paper prototyping (see Chapter 3 and Muller, 1991). Storyboards are created by developing a preliminary design from a sub-set of the usage scenarios gathered in Phase 1. Storyboards are sketches or mockup screens that show key steps in user system interaction. The analyst walks through the storyboard explaining what happens at each stage in terms of system functionality, and asks for the users' opinions. A more engaging variation is to ask the users to walk through the storyboard having been given a motivating scenario. User-led walkthroughs focus on the user interface requirements as well as system functionality, and uncover design problems when the users cannot predict what to do next or find the command they require.

The limitation of storyboards is their poor interactivity. Users can be asked to simulate the actions they would carry out, but mimicking the system response is more difficult. A set of output storyboard screens should be prepared to anticipate possible actions the user may take, so different pathways through the system can be followed. Inevitably there will be user actions that were not expected, but these are productive indications for missing requirements and design improvements. A storyboard for the shipboard emergency management system is illustrated in Figure 6.2. This gives a series of sketches illustrating the hazard detection and analysis scenario when fire breaks out in the cargo.

The key points to note in storyboard design are:

● The sequence illustrates key steps in the scenario, and each image/screen is linked to show the overall sequence.

● The storyboard steps are annotated with descriptions of system functionality, and possible design implications. These annotations act as an aide-memoire for the analyst when explaining the sequence. Further annotations can be added to record user comments about design features.

● In interactive storyboarding a set of different screens/images are manually presented to imitate alternative pathways through the dialogue.

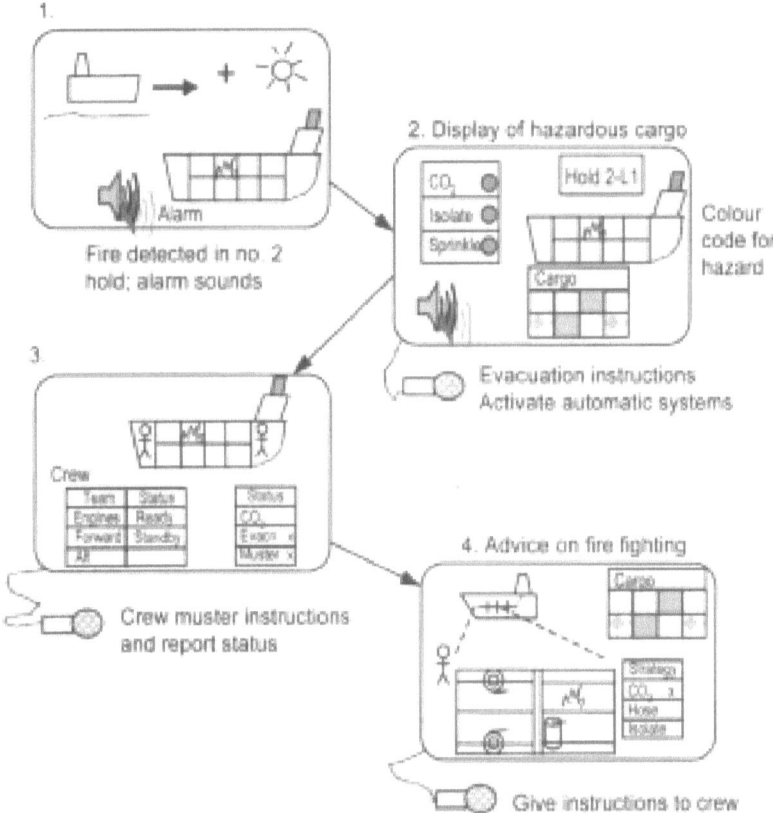

Fig. 6.2 Storyboard sketches for the ship emergency management system.

A combination of storyboards is often advisable. High-level overviews of the system functionality can be presented in an analyst-led walkthrough while areas of system detail are explored in user-led walkthroughs to test how well the early design ideas support the users' tasks.

Problems reported by users are recorded, as well as suggestions for improvements. One of the merits of storyboards and scenarios is that they help involve users in design. When users voice concerns about a design, the designers should not be defensive. Admit that this is only a first cut and then ask the users for their suggestions. Better still, provide pens and paper and make the session interactive and collaborative. This changes storyboards into paper prototypes and encourages user participation. The process is as follows:

- Prepare storyboards and scenarios. Provide blank screens (B4 sheets of paper), post-it notes and pens in different colours.
- Brief users about the scenario and storyboard, and how re-design can be done with paper prototyping materials.
- Walk through the scenario/storyboard, gathering users' reactions and suggestions for improvements. Redesign with paper prototyping when problems are encountered; alternatively, run through the whole storyboard first before engaging in redesign.
- Redesign the storyboard with users. The new design may then be re-run through the walkthrough, although a period of reflection is usually necessary to collate all the suggestions; refine the design and then return to the walkthrough in a subsequent session.

Storyboards and paper prototyping allow for quick iterations of a design, but they can mask the system functionality. Although some iterations of storyboarding will be necessary to refine the initial system vision and assess alternative views, it is advisable not to prolong storyboarding beyond two or three iterations. Better feedback will be obtained when a more interactive prototype can be demonstrated in the next phase of requirements exploration.

The outputs from this phase are:

- amended storyboards recording the preliminary design;
- a list of high-level requirements;
- recordings of user suggestions.

6.4 Requirements Exploration

This phase starts when the design vision has stabilized and sufficient analysis has been undertaken to build a scripted demonstrator to illustrate how the system will work with a limited set of scenarios. The additional task in this phase is to specify the system functions. SCRAM does not explicitly cover modelling and specification, as this is assumed to progress in parallel, following the software engineering method of the designer's choice (e.g. UML and unified process). At this stage some modelling and specification will have been undertaken so that system functionality can be simulated, although details of algorithms and data structures can be left until later in the process.

6.4.1 Session Design

The physical layout of the analysis session is illustrated in Figure 6.3. The setup is intended to encourage co-operative requirements capture between two, possibly three users and two requirements engineers. One acts as a driver of the concept demonstrator and the other fulfils an explainer-rapporteur role. The presence of at least two users helps balance the "ownership" of the session away from the developers and is productive in producing conversation about the artefact, domain and requirements. Larger user groups can be handled if a data projector is available to display the concept demonstrator; however, results are better with a group size below six. Larger groups slow the session and create difficulties in managing floor space for each participant to have his/her chance to voice an opinion.

Prior to the session the concept demonstrator is developed and tested. A representative set of users should be selected and divided into groups. A contextual scenario is developed based on the preliminary domain analysis. This is a short narrative (half to one page) describing a situation taken from the users' work context, e.g. "a typical day in the life of ..." running through key tasks. It should also contain sufficient background material to "situate" the action – that is, to give the users enough information to interpret the script. An example for a ship emergency management system follows.

6.4.2 Contextual Scenario

Your ship is a modern container ship of 30,000 tons displacement with a multinational (mainly Filipino) crew of 36 with 5 UK officers. The cargo manifest lists containers with a mixture of industrial goods and domestic removals. You have left Southampton at 10 a.m. en route to Cape Town and are proceeding west down the English Channel heading 250 degrees at 15 knots, position 50 miles south of Plymouth. The weather is fine, visibility range 15 miles, wind NW force 3. At 3.10 a member of the crew reports smoke in number two hold, and a fire alarm sounds.

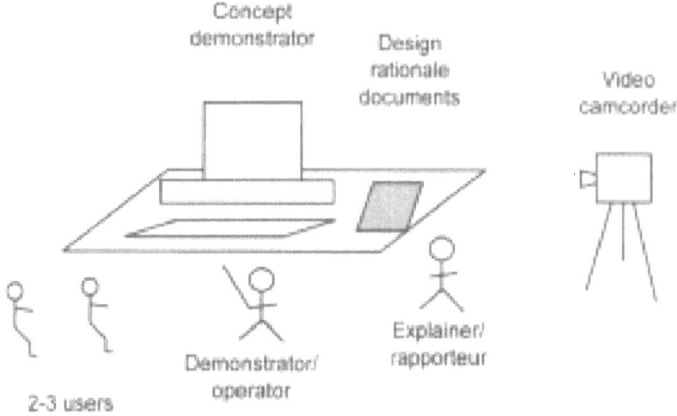

Fig. 6.3 Layout of the requirements analysis session.

A *scenario script*, based on the captain's emergency management task, is used for the concept demonstrator.

1. The location of the fire is investigated and preliminary instructions given to the crew to evacuate the area if necessary.
2. Automatic fire suppression systems are activated if present, such as flooding compartments with CO_2.
3. Instructions are given by tannoy to the crew to proceed to fire muster stations and start to fight the fire.
4. Junior officers are assigned to manage key fire fighting teams.
5. The cargo manifest is checked to see if any dangerous (explosive, flammable, corrosive, etc.) cargo is present near the fire.
6. Fire-fighting tactics are planned to account for any dangerous cargo and other hazards, e.g. electrical equipment.
7. Instructions are given to fire-fighting crews. The progress of fire fighting is monitored and further instructions given as necessary, until the fire is under control.

The scenarios are sent to the users beforehand for comments and to act as briefing documents for the requirements session. Note that the scenarios do not attempt to cover all aspects of the users' tasks or the situation; for instance the damage assessment phase is not described, nor is the means of communication between the captain and crew. These details will emerge during the demonstration.

This phase consists of four steps:

1. *Specification and development of the concept demonstrator.* A concept demonstrator is an early prototype with limited functionality and interactivity, so it can only be run as a "script". Scripts illustrate a scenario of typical user actions with effects mimicked by the designer. Concept demonstrators differ from prototypes in that only minimal functionality is implemented and the user cannot easily interact with the demonstrator. Typical implementation environments are multimedia authoring tools such as Macromedia Director, HyperCard or Visual Basic. Development time is in the order of 2–4 days for a moderately sized application.
2. *Requirements analysis: validation session.* An introduction and briefing puts the users at ease, explains the developer roles and emphasizes that it is the artefact, and not the users, that is on trial. The concept demonstrator is illustrated in a scripted sequence, linked to the grounding scenario. Probe questions are asked at key points in the demonstration script. The users are invited to critique the concept demonstrator. The concept demonstrator is explained by the analyst, while another member of the design team runs and interacts with the system. Critiques and suggestions for improvement are captured by notes and recording the session for subsequent analysis. Scenarios are used to situate the demonstration and discussion. Limited hands-on testing may be provided at the end of the session. In a follow-up phase, the users are encouraged to clarify any points they found ambiguous, go back to any parts of the demonstration, and elaborate further requirements. The requirements engineers may also follow up points raised or user comments made during the session.
3. *Session summary.* Once the demonstration has been completed, a summary of the requirements is listed on a whiteboard (or another appropriate medium).

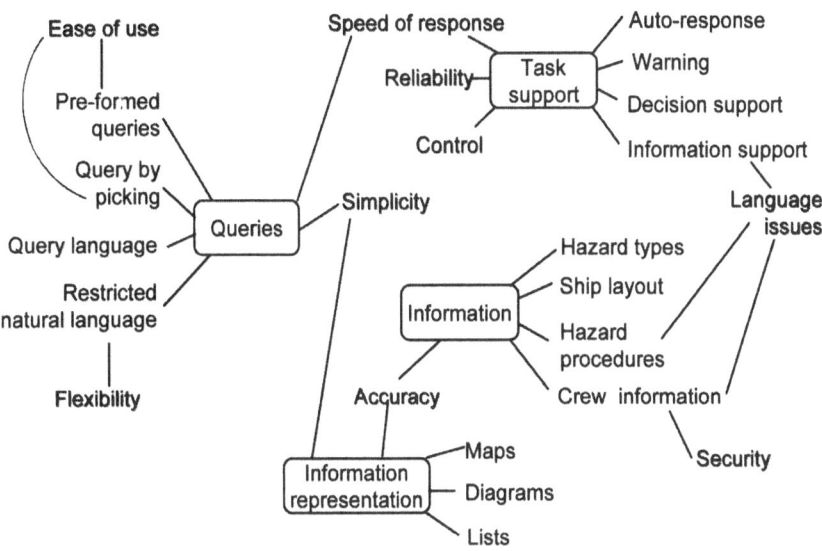

Fig. 6.4 Network map of requirements, showing relationships and groupings.

Requirements may be represented in a network map to show relationships and groupings (see Figure 6.4). The requirements are discussed and prioritized using an essential/useful/optional scale. Design rationale may be introduced in this step to help structure discussion about design trade-offs with assessment criteria (often non-functional requirements) that can be used to judge the merits of alternative solutions. It is useful to have the concept demonstrator present to refresh users' memory about each design option. In the summary phase, the explainer-rapporteur summarizes the key facts learned during the session and requests any comments. If the users wish to take copies of the concept demonstrator away with them they are encouraged to do so.

4. *Session analysis.* Data collected during the analysis session is analyzed and conclusions reported back to the users. This frequently leads to a further iteration of revising the concept demonstrator and another analysis session. Following the session the video data and audio soundtrack are analyzed in conjunction with notes taken during the session. The depth of analysis depends on resources available. A complete transcription may be undertaken so the users' verbal commentary can be semi-automatically analyzed for different requirements and domain facts. Alternatively the video may be "eye-balled" and facts analyzed by playback of key sequences.

The end point of this phase delivers a requirements specification comprising:

● the concept demonstrator;

● a set of requirements possibly supplemented by design rationale diagrams expressing users' preferences for different design options;

● specifications as text, graphics or more formal notations depending on the requirements engineer's choice.

In addition, video of the analysis sessions is available for subsequent requirements analysis.

6.5 Walkthrough Approach

SCRAM provides outline guidance on how to conduct a session, as well as heuristics for fact capturing and handling the user-analyst discourse.

6.5.1 Operational Guidance for Analysis Sessions

The session is started with an introduction and verbal summary of the situation described in the scenario narrative, e.g. "imagine you are in your office and production order arrives ...". One developer operates the concept demonstrator while the explainer-rapporteur asks questions at key points in the demonstration script. It is important for the developers not to dominate the floor space so users can give their opinions freely. Hence questioning is restricted to a small number of key points in the script. However, when users are shy or not forthcoming the explainer-rapporteur should take the initiative and prompt a user response.

Figure 6.5 illustrates a screen dump from a shipboard emergency management system. The user's requirement is for timely and appropriate information to support their decision making. The operational steps accompanying Figure 6.5 are:

User: identify the hazard location.
System: shows location of fire.
User: sound alarm.
User: find location of fire-fighting crews.
System: displays crew information and location on the diagram.
User: decide appropriate instructions to give to crew.
System: displays a checklist of actions.

The key point in the task is how to instruct the emergency team on where to go and how to deal with the hazard, in this case a fire. The concept demonstrator illustrates one design option. Alternative solutions expressed in design rationale format are

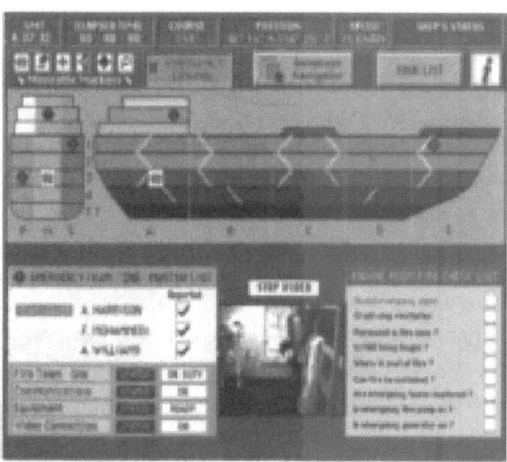

Fig. 6.5 Concept demonstrator showing the "show emergency teams and hazard location" design option for the "muster emergency teams" task.

Steps in scenario   Design rationale diagram

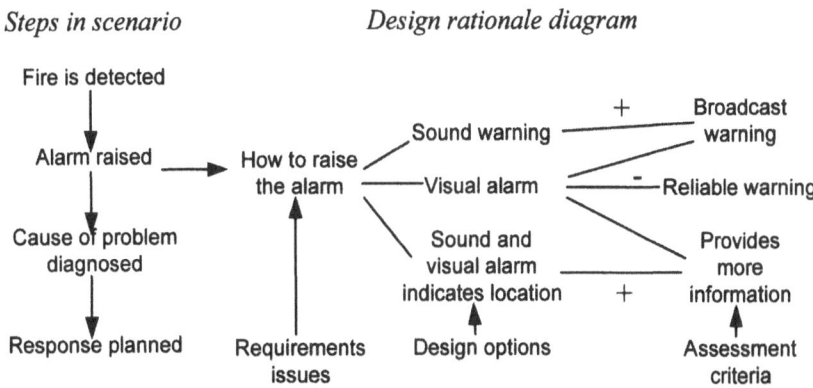

Fig. 6.6 Design rationale diagram showing the scenario script and design options for one step.

illustrated in Figure 6.6. The first option displaying comprehensive information is illustrated with the demonstrator, followed by option 2, provision of more restricted but relevant information for the task. This identifies the team nearest the fire, while the final option assumes emergency team autonomy and simply broadcasts the location of the fire. The users are asked to rate each option and consider the trade-off criteria. The design rationale diagram can provide a recording medium for ranking of options; furthermore, additional ideas and notes can be scribbled on top of the diagram. Generally the diagrams are more effective in the summary phase, as introducing them during the session distracts users from the concept demonstrator. Alternative designs are represented as storyboards with the best option in the concept demonstrator. However, should additional resources be available, alternative versions of the concept demonstrator can be implemented and both versions used to illustrate a key point.

The number of key points planned per session depends on the size of the application and number of requirement issues the developers wish to explore. As an approximate guide, sessions are advised to last no longer than 60 minutes (e.g. the straight demonstration script is 20–25 minutes) with 6–7 key points per session. Balance of questioning and conversation initiative is ultimately a matter of the experience and sensitivity of the requirements analyst.

6.5.2 Questioning Strategies

These can be regarded as side loops within the main thread of demonstration-led conversation. There are three patterns (see also Chapter 4):

Implication Exploring Questions

The questions amplify concepts drawn from the inquiry cycle (Potts *et al.*, 1995) and may be linked into key points in the script. Questions are prepared beforehand and focus on input and output events across the boundary of the intended system.

For inbound validation, the impact of missing, mis-timed and inappropriate events is analyzed to suggest system requirements to deal with missing or unexpected events, excessive volumes of input and dangerous events in certain contexts. Questions encourage the users to explain the range of different event types that might be input, likelihood of delayed or missing events, etc. For example: "What happens if the emergency warning is delayed?". When the developer discovers events which the current specification cannot deal with, follow-up questions are used to elicit what the system should, or should not, do in this circumstance. If no system function exists to deal with the particular event then a goal must be added to the requirements specification.

For outbound validation the focus is on the acceptability and impact of system output on users. The contextual scenario can be used as a basis, and the users are prompted to imagine permutations on this scene and the effect different environmental contexts (e.g. climate, location), or user roles may have on the acceptability and appropriateness of the system output. Questions focus on whether the information would be appropriate or acceptable to different user groups or if system output and decisions will be workable in the real world – for example, "Will detailed instructions be understood by the emergency crews?", "Can warnings be heard in all compartments?".

Fact Capture Discourse

Facts can be elicited either by semi-closed questions which contain some information indicating the expected answer, or by checking statements which test a user's belief (e.g. "So in normal operation distribute the hazardous cargo in different parts of the ship") accompanied by gesture to refer to the concept demonstrator. Fact capture may focus on gathering domain or task-related information, context data about the users, or further requirements. Initiative by the analyst should not dominate the session as users often volunteer facts in response to the concept demonstrator.

Intention Capture Discourse

This is mainly used in the summarization phase to elicit user requirements which have been omitted or are outside the scope of the current requirements modelling. Open questions are asked, for example "Are there further information needs?", and the scenario is used to critique the scope of the system, for example "Should the system cover additional areas of operation?".

Heuristics are provided to guide the questioning in different contexts. For instance, closed questions that require a yes/no answer are generally discouraged. Semi-closed questions are recommended for following up points raised during the demonstration, whereas more open questions are better for intention capture. Exchange structures (see Chapter 4) for user–analyst conversation are suggested as tools for thought for analysts to rehearse their explanations and practise questioning skills, for example:

Key point explanation pattern for the developer:

● Inform (design issue).

- Propose (Option 1).
- Illustrate (Option 1 on demonstration).
- Propose (Option 2).
- Explain (Option 2 narrative/storyboard).
- Justify (Option 1 with criterion 1), etc.

6.6 Session Summary

Requirements are summarized as lists on a whiteboard, flipchart or other medium. The requirements will be developed from the list recorded in Phase 2, with others captured during the demonstration session. The requirements list is structured into groups of functionally related requirements, or as a network of goals to show relationships, such as sub-goals, dependencies and conflicts.

The summarization session has three important objectives:

1. To discuss and agree alternative designs for particular requirements.
2. To clarify problems and ambiguities that arise during the demonstration session.
3. To agree a prioritization of the requirements.

The first two objectives are dealt with first. Decision tables or design rationale can help with discussion and prioritization using the following dialogue pattern:

- Introduce the design issue/problem.
- Explain each design option, and draw attention to evaluation criteria.
- Invite users to discuss each option in turn.
- Request prioritization of the users' favoured option, with voting if necessary.

The analyst then proceeds to the next requirement where alternatives need to be resolved. Discussion of alternatives may uncover assumptions as users start to unpack the design. This leads to decomposition of higher-level designs into more detail, which naturally suggest further requirements. Managing the level of detail in summary sessions presents a dilemma. On the one hand, the additional detail exposes further requirements but on the other hand, you don't want to become bogged down with excessive detail too early. Generally, mark points where decomposition became apparent and move on to the next requirement, to maintain a more consistent level of detail throughout the requirements documentation.

The choice of representation depends on the issues that need to be resolved. A hierarchical list of requirements can be used to structure the agenda of the meeting. Not all the requirements will need to be discussed, as some should have been agreed in early design visioning. The analyst needs to select the requirements that need further attention; for example, where disagreement was apparent between the users or where there was ambiguity. The users' opinion of discussion topics should also be sought. The design rationale diagram can promote discussion and illustrate differences between the options. One problem is bias towards the option implemented in the demonstrator. This can be counteracted by using storyboard sketches of the other options and by more vigorous critiquing of the implemented version by the developers. In particular, use of the criteria that incidentally capture non-functional requirements is a powerful way of promoting critical thought.

For general prioritization the dialogue management pattern is:

● Introduce the prioritization scheme and grading, e.g. essential, useful, optional.
● Draw attention to any key prioritization criteria. Cost is often the most important but there may be other non-functional requirements such as security, usability, etc. List these on a flipchart as an aide-memoire.
● Step through each requirement in turn asking the users to prioritize them.
● Rank the requirements in priority order and invite any further user comments. The first cut ordering frequently leads to some reassessment of priorities.

A frequently encountered problem with prioritization is users ranking everything as essential, i.e. the gold-plating option. To counter this give users a budget with a limited number of "essential counters" to spend; or, to make the session more fun, give them Monopoly money to the value of the development budget. Then ask users to price the options they prefer with their own money. This concentrates minds wonderfully.

6.7 Post-Session Analysis

Different levels of analysis can be applied to data captured during the session, depending on resources available and the fidelity of the required results. If resources are limited this step may be omitted, in which case the requirements captured during the session and the refined concept demonstrator become the output. The merit of further analysis lies in capturing user suggestions and design critiques that may have been missed during the demonstration and summary. Even with a team of analysts it is easy to miss some key points when concentrating on explaining a design.

The simplest analysis is to just eyeball the video recording and listen to the users' conversation to capture any points missing from the session notes. If the video and audio sound track have been transcribed, text-searching software can be used to create word lists to help fact capture. Simple filters can be devised to sort facts into categories. For example:

● *Domain facts*: nouns are good markers for domain facts and useful indicators for constructing entity relationship models, e.g. ship, compartment, cargo, hazard, crew, etc.
● *Functional requirements*: verbs serve as reasonable markers, e.g. locate (hazard), evacuate (crew); however, verbs also indicate task actions and events (e.g. send warning), so lists have to be scrutinized with care and keywords examined in the context where they occurred.
● *Non-functional requirements*: suffix searches (e.g. *ity, *ality, *ly, etc.) help to locate adverbs and participles which frequently express non-functional criteria such as quickly, safely, security.

In each case software can only help with pre-processing. Human expertise is required to extract reliable meaning from transcripts.

The output from this stage is:

● prioritized requirements list;
● analyzed session recordings, marked up transcripts and videos;
● concept demonstrators, refined in response to user comments.

6.8 Some Warnings

Even though the SCRAM approach is a powerful means of requirements analysis, it does have some dangers that practitioners should be aware of. The first problem is *bias*. Showing one design option in a concept demonstrator will tend to influence users in favour of that option over others that are demonstrated less completely. Ideally all options should receive equal implementation, but in practice resources are limited and there is a tendency to implement one option and show alternatives as storyboards or as simpler, less interactive displays.

The second problem is *session overloading*. Running concept demonstrators with scenarios can take considerable time. The length of scenarios should be limited to a maximum of 4–5 key points where design alternatives need to be explored. In practice this means that scenarios may have to be subdivided into separate sessions. The number of key points per scenario will depend on the complexity of the users' task and the design. Design rationale diagrams may be used as an aide-memoire for the analyst, rather than as a shared representation during the session. Too many representations during one session create problems of context switching for users when they have to transfer from the concrete concept demonstrator to more abstract concepts in design rationale diagrams.

Running a successful session is also dependent on thorough *training* of the requirements analyst. We found considerable differences in the quantity and quality of requirements captured between analysts who were experienced and novices in using the method (Sutcliffe and Ryan, 1998). Experience is necessary to manage requirements conversations, in particular to follow up questions and make sure users get sufficient floor space to give their feedback. Nevertheless, the method can be used successfully with a modest amount of training.

6.9 Prototyping and Requirements Validation

Prototypes are developed either from the concept demonstrator or from scratch depending on the language and tools used. For instance Visual Basic can be used for a seamless transition from scripted concept demonstrators to functional prototypes. With visually complex user interfaces it may be more efficient to develop concept demonstrators in an authoring tool (e.g. Director) and then start prototyping afresh with a different tool. However, some care should be exercised to transfer the look and feel of the concept demonstrator to the prototype, otherwise users will become distracted by style changes in the user interface.

The objective of this phase is to obtain more detailed user requirements and validation feedback on the design. Although the earlier phases of SCRAM elicit requirements quickly at a low cost, user reaction is stimulated more effectively by hands-on testing as explained in Chapter 2. At this stage requirements validation converges with usability testing, so feedback is gained on user interface defects as well as on functional requirements.

6.9.1 Session Design

The preparatory activities before the validation session are:

● *Specify scenarios of use for testing the prototypes.* These will be drawn from earlier phases of requirements analysis. The number and range of scenarios is not easy to judge. More scenarios give better test coverage, but at more expense. Generally the number of scenarios should be judged by how many can be completed within 60 minutes' testing, and chosen to cover key user tasks and as many exceptions as possible. Scripts need to be developed from the scenario to describe the users' actions.

● *Develop the prototype with functionality* that matches the set of user tasks and scenarios developed for earlier SCRAM stages. In early cycles of testing, broader but shallower coverage of prototypes is useful to home in on areas of requirements problems, whereas in later cycles more detailed feedback will be required, so more in-depth but narrow prototypes will be required. Generally IS requirements are less certain and more open to change, hence broader prototypes are better in information systems, whereas narrower prototypes with more functionality are better for complex systems where algorithms and controls need to be checked.

● *Select a representative set of users.* This is not as easy as it sounds because getting hold of potential users who are willing to give their time can be difficult; moreover, selecting an appropriate range of people from a potential user population may require many individuals. The issues to beware of are:

  - The *age and abilities range* of the user population. The sample should be balanced for age and gender. For office applications a sample of office personnel is necessary but for a general-purpose graphics package a wider sample of the public, young and old, will be necessary. Testing for a range of abilities and disabilities is becoming increasingly important for many products.

  - The *nationality and culture* of the potential user population. In international user interfaces, including many web applications, a range of linguistic abilities and cultural differences will need to be considered.

  - The *level of computer experience* and knowledge of the operating system. Younger users will tend to be more familiar with computers than older users.

  Some organizations have user panels that provide a ready-made sample for testing their products; however, there is a danger of using "captive" users too often. Tailoring the user sample to the expected population of the product is better. Generally more users gives better results, but there is a law of diminishing returns in requirements validation, and testing 8–10 users will usually discover most of the important requirements defects.

● *Prepare the test environment.* The sophistication of the test environment depends on the quality of data required. Usability laboratories evaluate products in special rooms equipped with video recorders and one-way mirrors to make observation as unobtrusive as possible. However, this approach can still be criticized because it places the user in an artificial setting. Testing in the user's workplace, in contrast, is more natural and helps put the user at ease (Beyer and Holtzblatt, 1998). Whatever the approach adopted, the location and recording equipment for the session need to be prepared.

Once the session has been prepared, requirements validation can commence. Users can be tested individually or in pairs. Testing pairs of users has the advantage that their conversation often reveals many problems without the need for follow-up questioning. Within each session the steps for requirements validation are as follows:

- Subjects are asked to complete a *pre-session questionnaire* about their computer experience (e.g. Microsoft Windows). This establishes a baseline of experience which is useful in judging individual differences.
- Users are *trained* with the product. The training should ensure that the users can operate the system sufficiently well so most can complete the evaluation. Training consists of demonstrating system facilities, then asking users to run through representative tasks (different from the test tasks) and answering problems. The amount of training depends on the type of product. For a general public "walk up and use" product little training should be given because it is important that people can use it without support. In contrast, a more complex office product which will be used by skilled users will require more training.
- Users are *briefed* about the tasks and reassured that it is the system that is being tested not them. Testing is hands-on without intervention from the analyst unless the user becomes totally stuck, in which case the reason why should be ascertained and help given.
- Users are asked to *complete the test* following the set tasks and to give a commentary when they experience difficulty with the system. Some users are better at verbalizing their opinions than others, so it helps to give examples of the feedback required during the training session. The analyst observes the users and if difficulties are apparent and the user does not give any reason, then a mini-interview is used to elicit the reasons for the problem. Users should work at their own pace and their behaviour and commentary may be video recorded if necessary. Problems encountered need to be noted by the analyst, with task completion times.
- A *de-briefing session* is carried out after the end of the test to follow up on the reasons for problems, missing requirements and to elicit suggestions for improvements. The de-briefing sessions provide an opportunity for the evaluator to go through any usability problems the user encountered.
- A *post-test questionnaire* can be added to capture user attitude on 1–7 scales. This provides more quantitative feedback on the users' reactions to particular design features and gives a population-level view of the users' responses to the prototype.

Each session should last around one and a half hours, but may take longer. The duration of the hands-on test session needs to be limited to about 45 minutes so the test tasks have to be selected with care.

6.9.2 Data Collection and Analysis

The objective is to identify missing or inadequate requirements and obtain feedback on user interface design features for improvement. The key outcome of validation is to link observed problems with the responsible design features and suggestions for improvement. Requirements validation is thus diagnostic or forensic in nature. In many cases the problem is obvious when the user reports it; however, some problems have deeper-seated causes that are not apparent to the user. In these cases the analyst has to diagnose the cause by observation and questioning. The first task is to categorize the observed problems. The first three categories are true requirements problems, while the remaining ones indicate user interface design defects:

Requirement Defects

● Task fit (missing functionality): the system does not contain a function for the user's goal, indicating missing requirements. Missing requirements are usually reported directly but may be indicated by the user searching unsuccessfully for a command or icon.
● Task fit (poor support): the system is used correctly but the user has to work around the system to achieve their goal and perform sub-optimal or redundant actions, indicating that support for the user's task could be improved.
● Task fit (inadequate functionality): users can partially achieve their goals but they find it difficult to do so and functional requirements do not exactly match their expectations.

User Interface Defects

● Cue/prompt/metaphor (poor location): the requirement has been implemented but the user interface features do not guide the user to find the appropriate command or manipulation.
● Cue/prompt/metaphor (predictivity): the interface does not help the user guess how to operate the system, or suggest possibilities for future actions. This problem is related to poor task fit.
● Manipulation/operation: cursor movement or other manipulations are difficult because the target is too small, or hard to operate, or carrying out the action exceeds the normal physical co-ordination abilities of the user.
● Missing feedback: no message or effect is visible or audible.
● Inadequate feedback: feedback is present but is either ambiguous or not sufficiently salient so the user overlooks it.
● Hidden effects: modes or parameter settings are not apparent or have been forgotten by the user, leading to unexpected effects.
● User error: the system is used correctly but sub-optimally for the task. This may be a training problem but task fit problems should also be considered.

The data available to derive these categories will be either verbal reports or problem observations. If video recordings have been taken these can be replayed by first eye-balling the session to spot potential problems, which are marked and then investigated in more depth. The prototype design feature (e.g. screen, command, menu, icon) being used when the problem occurred is noted. In the absence of an explicit user description of the problem cause, the following heuristics and decision tree (Figure 6.7) can guide analysis from observed problems to the potential cause:

● At the beginning of a task or sub-task, if the user cannot proceed or is puzzled, then suspect either missing functionality (requirement not implemented), hidden functionality error (can't find function/command), task fit error (missing functionality) or user error (poor task/domain knowledge).

● After a successful action, if the user cannot proceed or is puzzled, then suspect goal formation problem owing to either missing/inadequate functionality or task compatibility error.

● If the user has found a target object/command but hesitates or cannot proceed, then suspect cue/prompt/metaphor error because the action is not predictable.

● If the user has completed the action but is puzzled by an unexpected effect, then suspect either inadequate or absent feedback (the system is in a different state from what the user expects).

In many cases the source of the problem is obvious, for example an obscure menu name, ambiguous icon or incomprehensible feedback message; however, in other cases the system behaviour appears bizarre to the user but the culpable design feature is not apparent. These errors are often caused by parameter settings in the system which are not obvious to the user (e.g. hidden broadcast options in e-mail programs), or by inadequate task models so that the system behaves in a manner that the user cannot comprehend – for example the legion of cut and paste (special), and insert options available in Microsoft Word which all have unpredictable and obscure effects.

The requirements and usability problems are analyzed to create a matrix of problem categories by design feature and missing requirements. The frequency of the

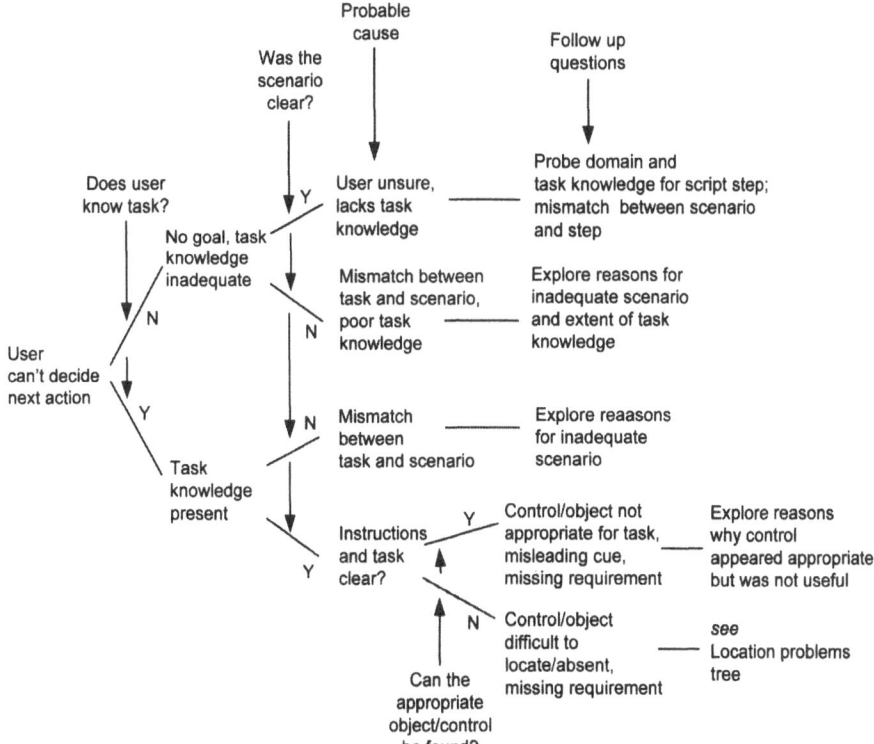

Fig. 6.7 Diagnostic decision tree.

problems is used to prioritize their severity and allocate resources to fixing them in the next prototype version. Missing requirements and design features which give frequent errors for all users are serious and must be fixed; however, features which cause all users problems, but infrequently, could have easily learned cures and probably require no modification. Problems encountered by one or two users are given a low priority for improvement. Quantitative analysis produces a ranked list of serious problems that merit further investigation.

Errors are categorized according to their origins in the interface design or user task/system knowledge:

- application system errors, sub-typed by cause (prompts, feedback, functionality, task fit, manipulation, etc.);
- operating system errors, including hardware and network problems;
- user errors caused by lack of task or operating system knowledge.

Operating system errors cannot be immediately changed, so training or changing the system configuration to avoid the problem are the only remedial treatments. This policy may also be adopted for errors caused by legacy systems; for example, backward compatibility with previous versions. User errors are investigated to discover missing or incorrect requirements. The outcome of this stage is a set of ranked requirements and usability defects, which identify missing requirements or defective design features. The prototype is modified to take account of the problems discovered in the first round of testing and then evaluated for a second time. The observed problems and missing requirements should have decreased fairly dramatically. If this is not the case then the user population and initial requirements are suspect. When there are two sets of stakeholders with conflicting requirements, satisfying one group may only upset the other. In most cases the prototype will improve to a satisfactory level after 3–4 cycles of testing although this will depend on the complexity of the product and the quantity of modifications made in each cycle.

This is the final stage in SCRAM, which should produce:

- validated prototype;
- requirements documentation, with prioritization;
- list of design defects that have been remedied in the prototype or need further attention in the final development.

6.10 Postscript

You may have noticed that models and specifications have not appeared in SCRAM. This aspect of RE is important but the omission is deliberate. SCRAM is intended to be a complementary method that can be interleaved with standard modelling/specification/design approaches such as object-oriented analysis, UML, and so on. The reason for proposing a separate strand for requirements analysis is to help communication between users and software engineers. SCRAM gives a process and representations for effective communication between users and analysts, a lingua franca based on combinations of interactive representations of concrete artefacts and scenarios. Hence from the users' viewpoint, requirements become real and tangible while being set in a context of use with scenarios. Abstraction is limited to trade-off decisions between different design options and prioritizing requirements. In the parallel technical analysis strand, abstract models and specification form the common language for software engineers. Requirements engineers act as the translators between users on the one hand and designers on the other, and need to be able to translate between concrete and abstract representations.

6.11 Summary

This chapter described a practical method of scenario-based requirements analysis. The method starts with initial fact capture to scope the vision of the intended system. This is followed by storyboarding and design mock-ups to get early feedback from users on preliminary design visions. The next stage develops the design as a scripted concept demonstrator that is evaluated with users by running through a scenario script, explaining design alternatives and asking users for feedback. Concept demonstrators refine the design and produce a list of detailed requirements. The final stage is to develop a functional prototype for hands-on user testing. The method guides analysis of missing and inadequate requirements as well as diagnosis of usability defects. Advice is provided on session management, handling dialogues and analyzing recorded data. Checklists and decision trees help categorization of requirements problems and usability defects. The prototype is improved and tested in a cycle until the requirements stabilize, after which the final product is developed.

7 Requirements Analysis for Safety Critical Systems

Requirements not only have to deal with events that are unexpected, they also need to anticipate the consequences of things going wrong. The unexpected can arise from many sources. Environmental events, often attributed incorrectly to "acts of God", create problems through adverse weather and physical conditions, causing machinery to break down so that normal system functions cannot be assumed. Furthermore, people make mistakes. Even though many accidents are blamed on people making mistakes, human error is usually only one of many contributing factors. Safety critical systems should prevent, or at least reduce the chance of, human error. Requirements analysis has to try to anticipate these problems, but this is a difficult task. The problem is one of 20/20 foresight, or trying to anticipate all the possible combinations of future events that might occur in a designed system, and then trying to ensure that the design prevents or at least counteracts possible failures. Given the large number of possible errors people can make, in combination with extremes in weather and the many components in complex systems that could fail, we have a combinatorial explosion. So anticipating the future is a very difficult task; however, that does not mean we should not attempt it. Even standard methods such as object-oriented analysis and use cases draw attention to alternative as well as normal courses of action.

This chapter explores alternative courses in scenarios, the reasons for human error and system failure, and looks at how potential risks of failure can be anticipated during requirements analysis. Failure is a problem in most systems, but in safety critical systems it can be a life-threatening concern.

7.1 Problems and Issues in Safety Critical Systems

Computers in safety critical systems usually fulfil two roles: monitoring and controlling an external system. Requirements therefore have to focus on how changes in the external world can be detected and interpreted, and then how the system can be kept within the bounds of safe operation. However, as complex systems cannot be controlled with 100 per cent confidence, the possibility that the system may depart from a safe state has to be anticipated. This brings in several new sets of requirements that more conventional systems do not need to worry about:

- requirements to detect unsafe states – monitoring;
- requirements for diagnostic support – how to interpret the cause of the problem;
- requirements to contain the immediate consequences of failure and protect people – e.g. evacuation procedures, shutting down damaged systems, fail-safe processes;
- recommendations for returning the system to a safe state, including repair of damaged components;

149

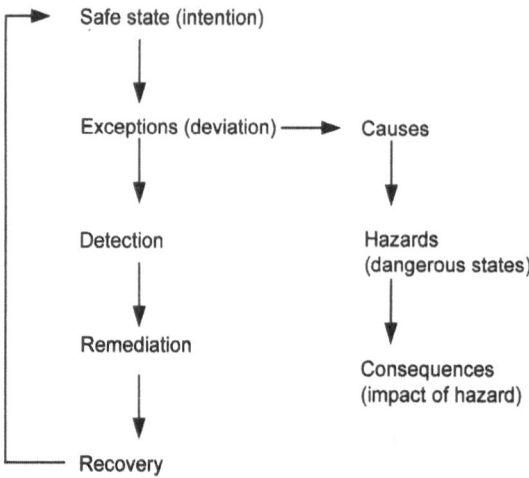

Fig. 7.1 Safety critical system control loop, with sources of error emanating from people, social setting, environmental conditions and component failure.

● methods for dealing with the longer-term consequences of failure: injury, environmental pollution.

The relationship between the computerized system and the external system it controls, with potential sources of error, are summarized in Figure 7.1.

Failure can have many causes. Indeed, as Reason (1990) points out, most accidents have multiple causes that arise from background conditions that allow the accident to happen, and a combination of events which the designer never anticipated. The "window of opportunity" increases the probability of failure by poor maintenance, inadequate training or poor motivation in operating staff. Poor design also contributes when requirements for the user-system interface make errors more likely and hinder diagnosis when things go wrong.

Anticipating errors, such as those in the A320 airbus described in Chapter 1, implies that requirements need to be investigated on several levels, starting with the social environment and system context which would not receive detailed attention in non-safety critical systems. We shall consider taxonomies of potential failure causes, but before that we need to investigate human error a little more closely.

7.2 Analyzing Requirements for Dependable Systems

Hollnagel (1993) proposed a useful distinction between events which are observable failures leading to accidents, which he called *phenotypes*; and the underlying causes of failure, called *genotypes*. Hollnagel's taxonomy of phenotypes in a simplified form suggested the following departures from normal behaviour:

● *event timing* – events might be too early or too late;
● *event patterns* – events may not occur at all (omissions), or may occur in the wrong place (commission, insertions, duplicate events, reversed ordering);
● *event quality* – the information contained in the event is corrupt or inadequate.

Requirements analysis has to anticipate the impact of possible event phenotypes and specify defences against them. Failure genotypes are often multifaceted because several influencing factors may be present that allow a rogue event (the phenotype) to occur. Hence the analyst needs to specify acceptable levels of requirements for defences against event patterns input from the system environment, including mistakes by people. Requirements for influencing factors concern the social systems rather than the technical system, and are frequently non-functional in nature, that is, targets and levels of service that must be maintained. However, some have more direct requirements implications for personnel training and technical system requirements. The following sections first deal with influencing factors, followed by analysis of event genotypes.

7.2.1 Influencing Factors

A taxonomy of *influencing factors* helps in understanding the potential sources of error and discovering safety requirements. The taxonomy and accompanying analytic techniques are based on research in the EU CREWS project SAVRE (Scenarios for Analysis and Validation of REquirements) (Sutcliffe, 1998; Sutcliffe *et al.*, 1998). It provides an explicit analysis of design requirements arising from the consequences of failure and human error; high-level guidance on reasons why errors might occur; and suggests generic requirements to deal with the causes and consequences of error. Generic requirements are reusable heuristics for preventing or ameliorating potential system failures and can be used as an agenda of issues pointing towards areas for further investigation; alternatively, the requirements that can be refined with more domain-specific knowledge, into design solutions that may be added directly to a requirements specification.

These influencing factors are grouped into four categories:

1. management and organizational;
2. user/personnel;
3. environmental;
4. task/domain.

Each affects different human internal variables such as fatigue, stress, workload and motivation. These, in turn, affect the probability of human errors, which are manifest as slips and mistakes.

Management and Organizational Factors

This stage focuses on awareness of safety, discipline in operating procedures, effectiveness of training, and workforce motivation. Managerial initiatives and control procedures are assessed in light of the demands of the task, operator competence, and operational environment. Attention is also paid to maintenance procedures to ensure equipment is safe. Poor maintenance is a frequent cause of failure, when the original designer's instructions about how the system should be maintained are neglected (Reason, 1997). Unfortunately maintenance procedures are not always followed even in safety-aware organizations.

Problems at the managerial and organizational level are often key contributions to system failure, although they are rarely manifest as the immediate reason.

Instead, managerial factors create a climate in which other contributing factors can flourish, leading to the "window of opportunity" or the unsafe state that allows an unexpected event to precipitate failure. Requirements analysis at this level produces recommendations for the social rather than the technical system. Some of the more common problems are summarized in Table 7.1. Even though procedures may be well designed, errors may occur if they are not followed, either because of poor staff training, lack of motivation or inadequate monitoring in operational, maintenance and quality assurance procedures. Misunderstandings in human communication are common; furthermore, necessary communication channels may be absent or inadequate. Poor management culture may be the result of an excessively authoritarian attitude or, at the other extreme, a laissez-faire approach. Staff may be demotivated due to a variety of social factors such as lack of recognition of achievements, lack of promotion, and job insecurity. The generic requirements to prevent these problems are summarized in Table 7.1 and elaborated in most books on human resource management (e.g. Hendry, 1995).

User/Personnel Factors

The implications of user factors are summarized in Table 7.2. Some of these factors can be measured objectively by using psychological questionnaires (Cacioppo, Petty and Kao, 1984). For instance, general ability and accuracy/concentration can be measured by intelligence aptitude scales, decision making and judgement by locus of control scales, while domain and task knowledge can be measured by creating simple tests for a specific task/domain. The main implications of the personnel factors are for personnel selection and training, while the generic requirements indicate the need for computer-based intelligent assistants, critics, and aide-memoire information displays.

Table 7.1 Management and organizational factors that lead to a window of opportunity for error and system failure.

Managerial pathology	Initial effects/ manifestations	Secondary effects	Implications/ generic requirements
Authoritarian style	Poor staff motivation Lack of initiative Culture of fear	Inability to deal with unexpected events	Appropriate delegation for task/job description
Laissez-faire style without controls	Inconsistent processes Poor discipline	Incomplete tasks, mistakes and slips	Improve training on the job; scenario-based training
Poor procedural discipline	Task failure Degraded performance	Mistakes and slips, poor performance	Improve training, motivation; mentors
Inadequate communication	Misunderstanding Poorly co-ordinated actions	More mistakes, wrong decisions	Improve communication channels and technology, train staff
Inadequate resources	Stress and fatigue from increased time pressure	More slips, mistakes, poor decisions, poor checking	Increase personnel or supporting resources, or decrease job demands
Motivation	More slips Slow operation Short cuts	Increased mistakes, slow performance, short cuts	Improve incentives, job satisfaction, select appropriate personnel

Table 7.2 User/personnel factors with problems for slip- and mistake-type errors and generic requirements to counteract these problems.

Personnel qualities	Skilled task problems	Decision/problem solving tasks	Implications/ generic requirements
General ability	More slips, especially complex tasks Longer learning time	Inability to deal with unexpected events	Personnel selection appropriate for task/ job description
Knowledge of domain	More slips Longer learning time	Slow performance Failure in complex problems	Improve training on the job experience, scenario-based training
Skill/task knowledge training	Slow operation More slips	Mistakes, slow performance	Improve training, manuals, aide-memoires, mentors
Judgement/ decision-making	–	Poor initiative More mistakes Wrong decisions	Select personnel appropriate for task
Concentration/ accuracy	More slips Mal-ordered events	More mistakes Poor decisions Poor checking	Discipline and training to improve performance, select appropriate personnel
Motivation	More slips Slow operation Short cuts	Increase mistakes Slow performance Short cuts	Improve incentives, job satisfaction, select appropriate personnel

The impact of poor personnel factors for skilled performance and action slip-errors are shown in column 2, while for tasks that require decisions and problem solving, the implications for performance and mistakes are illustrated in column 3, with generic requirements and recommendations for training in column 4. User/personnel factors are used with assessment of task complexity (see Table 7.6), to match users' abilities to tasks of appropriate complexity, while also giving people sufficient challenge and responsibility in their jobs.

Environmental Conditions

Environment analysis focuses on the unlikely events that the system may have to deal with *in extremis*. Safety critical systems are often vulnerable to variation in climate and unanticipated effects of a local context. For instance, low temperature in an unusually cold Florida winter was a contributing factor to the *Challenger* disaster (Reason, 1997). A system may be designed for a set of tolerances but then moved to another location where these assumptions are violated. Requirements analysis should investigate the unexpected consequences on human and machine operation of excessive temperature, humidity, etc. The following heuristics link environmental factors with possible effects on human operator and machine reliability:

● If *ambient temperature* is too high/low then machine reliability may decrease, and human fatigue/stress will be increased.

● If *humidity* exceeds the expected range then there may be an effect on operator and machine reliability, with an increased tendency for human fatigue.

- If *noise levels* exceed a tolerable ergonomic range then there will be an adverse effect on operator performance, increased fatigue, and interference with verbal communication.
- If *vibration* exceeds the expected range then there may be an adverse effect on operator and machine reliability; instrument readings and human control manipulations will become more difficult to perform accurately.
- When *dust or dirt* are present machine operation may be compromised and require increased maintenance; human operator working conditions will be adversely affected leading to increased fatigue and stress.

Tolerance limits for the above environmental factors should be set for safe operation; however, even the best-intentioned design can be compromised unintentionally. Change in the operating location of mobile systems is one cause to consider:

- Will the system be used in different organizations?
- Will the system operators be changed?
- Will the system be moved to a new location?

Variations in environmental factors increase the probability of error, and may be quantified if sufficient historical data exists to propose baseline safety levels.

The implications for environmental factors, such as adverse temperatures, humidity, dust levels and so forth, on human errors and system safety are summarized in Table 7.3. Adverse environmental factors increase discomfort and stress for users, and these in turn will lead to more skill-based slip-errors when the operator's concentration fails. Moreover, when human operation involves decision making or problem solving, adverse environmental factors will lead to more mistakes and degraded performance.

Table 7.3 Environmental factors.

Factor	Effect on user/ operator	Potential errors: users	System failures	Generic requirements
Temperature too high/too low	Fatigue, discomfort, stress	Concentration fails, attention slips	Failure or degraded action	Air conditioning
Humidity excessive	Fatigue, discomfort, stress	Concentration slips, attention fails	Failure or degraded action	Air conditioning
Poor visibility	Eye strain, fatigue, stress	Slips reading data, transcription errors	Applies to optical scanners	Light screens, blinds, better VDU luminance
High noise levels	Audio communication difficult, stress	Misunderstanding, speech messages not heard, slips	Audio output interference	Sound insulation, noise protection for users
Excessive vibration	Visual communication impaired, stress	Misreading of visual displays, slips	Failure if not robust	Isolate and insulate from vibration
Dust and dirt	Discomfort, poor motivation, visual communication hindered	Concentration fails, slips	Degraded action, mechanical failure	Controlled environment, sealed under protective clothing

Task/Domain Factors

Implications for task/domain factors for user performance and errors are listed in column 2 of Table 7.4, with generic requirements and user training recommendations in column 3. Increased interruptions and task switching make slip-type errors more likely; higher volumes and time pressures also lead to more slip errors and delays. Complexity and repetitiveness, on the other hand, have implications for function allocation and increased mistakes if users are not given tasks that match their abilities, or training to carry out their allocated tasks. It should be noted that many task performance problems also have "knock-on" effects on users through increased levels of stress and fatigue.

Taxonomic analysis can point the analyst towards potential problems in different aspects of a system. The next step is to discover critical problems so that requirements to prevent them can be specified. To address this problem a method for modelling combinations of influencing factors is proposed.

7.2.2. Safety Critical Scenario Analysis

Influencing factors provide the conditions for errors and the window of opportunity for hazards. First, a scenario analysis is undertaken to test possible causes of failure and plan requirements for their prevention, containment or redemption. This can be followed by a more detailed, systematic scenario analysis at the event level, described in Section 7.2.3.

Scenarios can be acquired from accident histories and past experience within similar systems; but these scenarios will inevitably miss the unexpected event that

Table 7.4 Task/domain factors with implications for errors and generic requirements.

Task factor	Implications	Generic requirements
High volume	Delays, bottlenecks, performance degrades, fatigue	Buffer and smoothing workloads, automate if possible, design breakpoints, batch schedules
Complexity	User fatigue and stress, mistakes, problem solving failure	Match complexity to user abilities and experience, decompose task, simplify procedures
Repetitiveness	Boredom, poor attention, slips, missed events	Automate if possible, provide task variety, swap operators frequently
Interruptions	Attention slips, missing and mal-ordered events, capture errors	Provide aide-memoires, agendas, status/progress checks; screen-out unwanted events
Time pressure	Late events, slips, capture errors, omissions	Smooth event arrival, automate task, give user time to think, provide holding actions
Multi-tasking/task switching	Attention slips, missing and mal-ordered events, losing thread problems	Aide-memoires, agenda managers, clear mode/status indicators, schedule switching if possible

may cause the accident. Also accidents have an unpleasant habit of being unique (Perrow, 1984), so generalizing experience from similar systems is difficult. Of course lessons should be learned, but scenario analysis needs to go further and attempt to anticipate future hazards, so requirements can be specified to deal with them. Anticipating the future with 20/20 foresight may be over-ambitious, but scenario analysis can, at least, improve on no analysis. This section extends common practice in safety critical engineering where methods such as HAZOPs system modelling (Chemical Industries Association, 1993) and fault/event tree analysis (Swain and Weston, 1988) help to inquire about potential failures. The extension is to propose a more targeted set of questions with follow-up actions.

Scenario analysis, task and system modelling proceed in an integrated manner (Figure 7.2). The system model represents the external system that is controlled or monitored by the computer system for which requirements are being analyzed. The operator's task model describes how people will operate the computerized control system, and may be based on existing manual or software controls. The operator's task and system model are used to suggest scenarios of failure that are composed of the triggering event, the window of opportunity that makes the event potentially dangerous, and a description of the immediate consequences of system failure. The scenario is then developed during the analysis to discover requirements for preventing the failure in the first place, and if this is not always possible, the procedures for containing the situation and recovering the system to a safe state. This adds requirements for diagnosis and recovery tasks to deal with exceptions in the controlled system.

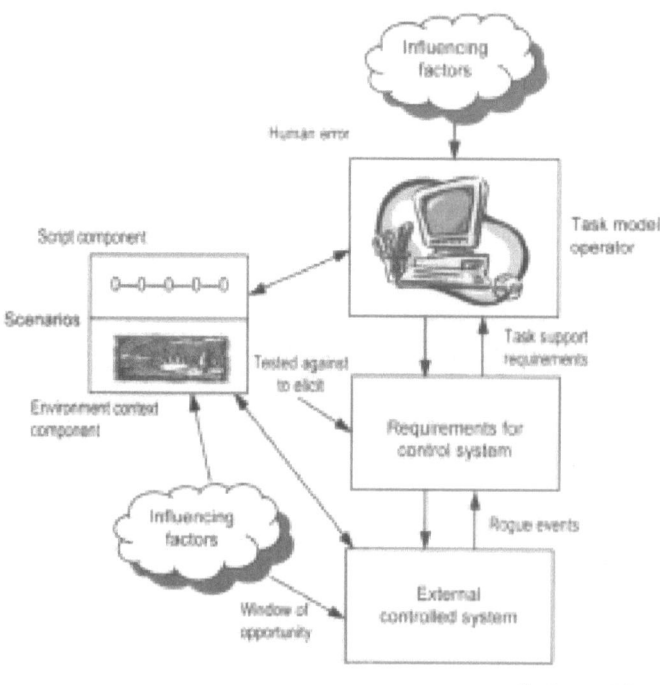

Fig. 7.2 Relationships between system models, scenarios and requirements in safety critical systems analysis.

Scenario Elicitation/Generation

Scenarios may be available from records of past system operation and these are invaluable in planning defences against future system failures; however, frequently the designer has few scenarios to work with. In this case, possible failures can be identified by examining the inter-relationships between the physical system model and the task model.

Scenarios have two main components:

1. *Environment context* describes the situation or window of opportunity leading up to the failure. The context component describes assumptions about the system-influencing factors, and should include situations when these depart from the ideal.
2. *Event sequence* and the immediate consequences.

Variations of environment and event sequences can be "played" in different combinations to expose problems when expected events become dangerous in a particular context. Environment components test the expected range from the normal case to worst case of potentially adverse parameters. Environmental variables can frequently be based on observation of the current system. The worst-case scenario should assume that all the variables are adverse and examine how they may interact. For instance, change in location may lead to exposure to excessive humidity, temperature and dirt; while poor maintenance of the equipment in the surrounding environment may also lead to excessive temperature and contamination. Environmental scenarios provide a background window of opportunity, or stress, that makes failures in equipment and human operation more likely. The impact of such hazards on the user's task are investigated, to create a scenario composed of:

- *event genesis* – the rogue event that causes the problem;
- *event detection* – when the user notices (fails to notice) something is wrong;
- *context* – the system state that makes the event dangerous;
- *consequences* of failure in terms of damage to the system, human injury, environmental pollution, etc.

The scenario will be expanded via questions to trace the sequence of events from failure through to recovery, in the following tasks:

- *diagnosis* – the user (mis)understands the reason for the abnormal event;
- *remediation* – corrective action is taken to contain the hazard and return the system to a safe state;
- *recovery* – any consequences of the hazard are dealt with and normal system operation continues.

Many different scenarios can be produced, so the problem is which ones to concentrate on. The system model is the first source of events that lead to failure scenarios. This follows a HAZOPs-like procedure of asking questions about what could fail in the physical controlled system, then creating the scenario of how the controlling system should respond to it. The scenario represents an anticipated sequence of events that is used to elaborate the requirements for the control system. In later stages of analysis other origins of failure events are suggested, leading to further scenario creation.

A case study is used to illustrate application of the method: a control system for a laser spectrophotometer. For this case study the task analysis, hazard analysis and gathering of domain facts and scenarios is reported in detail in Sutcliffe (1998).

The laser spectrophotometer is a scientific instrument that analyzes the visible spectra that are created by the laser ionization of a chemical sample. In normal operation the laser should emit a high-energy light beam that strikes the chemical sample causing it to emit light energy that is detected by the sensor. The light spectrum is analyzed and results are displayed on the computer VDU. The spectra have a characteristic pattern for each chemical. The physical system model is illustrated in Figure 7.3. Each laser emission cycle is controlled by software. There are two subsystems: one controls the operation of the laser, while the second detects and analyzes emitted spectra. The controlled system operational sequence is shown in Figure 7.3 below the model.

There are three types of system user:

1. *Expert users* – usually research scientists who possess considerable domain expertise and become skilled in system operation, i.e. task knowledge. They conduct complex analyses and require full control of the system for different analyses, with presentation of complex results. The experts' safety awareness is variable as not all laboratories have good safety practice.

2. *Chief technicians* – also expert users, who plan analyses in public laboratories and companies. Accuracy and reliability of the results are at a premium. These users need support for planning sessions and investigation of the results. Their safety awareness is generally good.

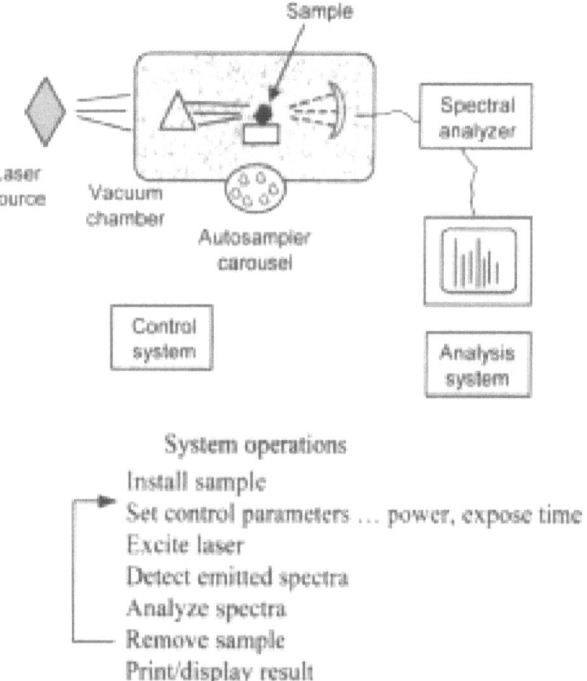

Fig. 7.3 Physical model of the spectrophotometer system, plus operational sequence.

3. *Skilled laboratory technicians* – whose knowledge of the domain and system operation will be average or low. From their viewpoint, system operation is a routine job. In multiple runs they may make errors from lack of attention and boredom. Their safety awareness depends on training.

The case study focuses solely on the third user group, hereafter referred to as the operator, and the activity of planning a sample run which is shown in Figure 7.4.

Task analysis identifies the normal operation and various fault scenarios, such as if there should be a power failure or failures in the emission detection apparatus (Sutcliffe, 1998). Environment analysis suggested a sensitivity to temperature, humidity and dust levels, which is of particular relevance given that these machines are intended not only for the UK market, but also for export around the globe. Excessive temperatures were found to lead to inaccurate readings and increased operator fatigue as calibrations then have to be carried out more frequently. High humidity increases condensation, leading to possible sample contamination and hence inaccurate readings. High levels of dirt or dust have the same effect on

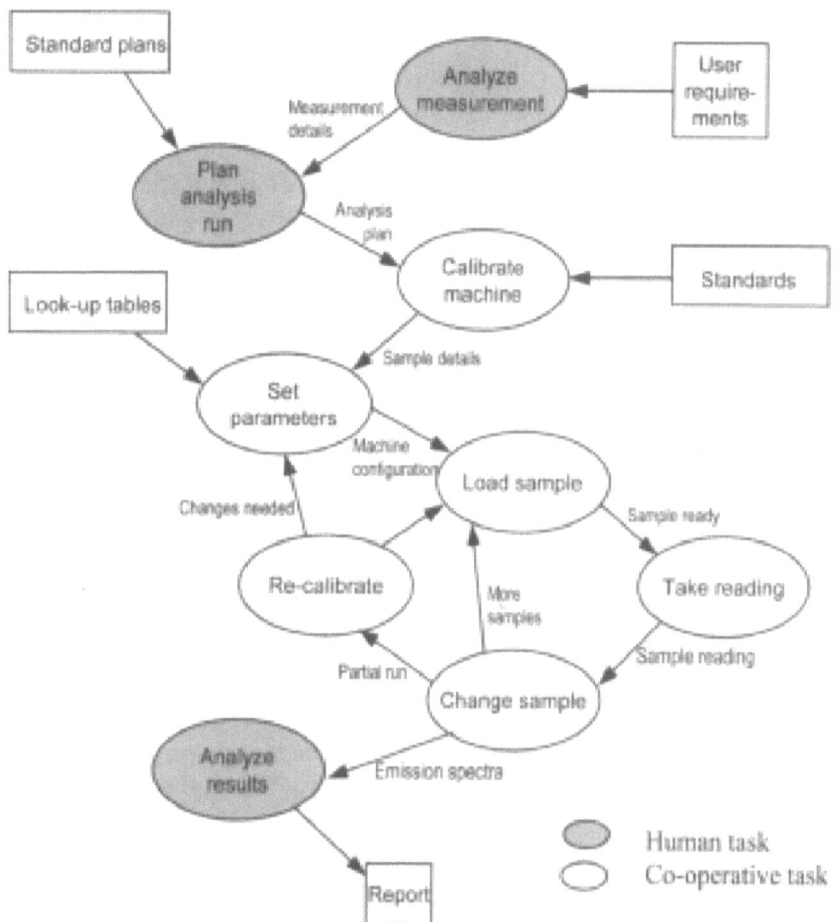

Fig. 7.4 Task model for the system operator expressed in data flow diagram format.

machine reliability and cause increased operator workload due to corrections and abandoned runs. Safety culture and level of operator training may also vary according to the place of operation. The worst-case environment is a poorly maintained laboratory with excessive ranges of temperature, humidity and contaminants, and no maintenance being carried out.

Functional Allocation

The scenario, task and system models are examined to establish requirements for computerized support of the operator's control tasks. A key question is the degree of automation in the response. This revisits the issue of functional allocation discussed in Chapter 3. Deciding whether people or machines should respond to warnings and dangerous situations will be determined by the same guidelines, although in safety critical systems the requirements analyst also has to consider the time available and the severity of impact. If the scenario analysis suggests that there is little time to respond and a severe consequence, automated response may be the only choice. The following heuristics (Leveson, 1995) guide the allocation of operator versus computer system responsibility:

● Match control tasks to operators' abilities and training.
● Automate monitoring and warning processes when possible.
● Provide decision support for exceptions/problems: causation analysis.
● Automate remediation/containment processes; give operators time to think.
● Leave people in control where possible.
● Make dangerous actions very difficult.
● Train with hazard scenarios and reuse experience.
● Promote a dynamic mental model of the system for operator awareness.

Scenario Event Analysis

Each step in the scenario is checked against the system requirements to ensure that there are functions in the controlling system to deal with event detection, decision support to help the user understand the implications, and control actions to deal with the hazard and take remedial action. Functional requirements will be identified by analysis of dependencies between the (assumed human) operator of the controlling system and the controlled system.

Analysis starts with questions about how the system will detect the event and introduces requirements for sensors and monitoring devices. The scenario is then elaborated with the human operator receiving the event, which raises requirements for display of error/warning messages. The next step is to describe the operator's action in diagnosing the problem, and this leads to questions about system support for this task.

In safety critical systems, information is vital to keep the operator aware of the controlled system's normal functioning and to warn about any exceptions that may require action. The scenario and task model should be investigated to answer the following questions for each event/action step.

Table 7.5 Questions to identify exceptions, with corresponding generic requirements.

Questions	Implications	Generic requirements
What types of failure are possible in normal operation, and when?	The quantity and quality of information, and the timing delay between the onset of a system change and its detection, need to be considered. Changes should be detected as early as possible and as quickly as possible, although this may not be feasible when the controlled system is remote (e.g. teleoperation). The requirements issues here are the detectability of events when not all the event types can be anticipated; also, some events may have no consequence without a dangerous pre-condition in the window of opportunity.	For sensors: ability to tune sensors, event pattern analyzers, sensors that can discriminate significant events from noise, and event logs.
How are failure events detected and signalled to the human operator?	Requirements issues are faithful detection of events, while preventing false alarms; also HCI issues of how to warn operators.	As for sensor/monitor processes, plus warning mechanisms, use of alarm sounds, voice, and visual highlighting techniques.
Can the event be understood, and its implications?	Situation awareness of the operator is an issue; does the operator have an up-to-date mental model of the system/domain, is there time to reason about the event, can jumping to false conclusions be avoided?	For warning messages: warning present in context of a system model or simulation, explanation of event, links to events and hazard implications.
What warnings (advisory) need to be communicated?	Deciding which events are potentially dangerous can be driven by safety case definitions of acceptable ranges for controlled system variables, e.g. the acceptable operating temperature and pressure in a pressurized water reactor. Tuning warnings can be more difficult. Sensors may give false alarms which may subsequently cause operators to ignore a real danger.	For sensor event interpreters: warning monitors for sensors, ability to tune warning levels from analysis of event histories.

Table 7.5 *Continued*

Questions	Implications	Generic requirements
What alarms (mandatory) need to be communicated? Mandatory alarms are defined by state variables exceeding the safety envelope.	The problem is setting the levels between advisory and alarms are used too frequently then operators can build up skilled procedure responses that they are hardly aware of. An example from the railway industry is the mandatory alarm when a yellow signal is passed. Drivers have to cancel a warning buzzer, which they do automatically, almost without noticing it. This has led to drivers running red lights.	For warning signals: user response dialogue, monitors on operator action (inappropriate response).
How can the failures be diagnosed?	Diagnosis sub-tasks may need to be specified for exception events which are not immediately obvious to the operator. Implementation of diagnosis may be eventually automated or handled co-operatively with the computer control system.	For diagnostic task support: locating the problem, gathering information, causal analysis, suggesting remedial action.
What diagnosis information should be displayed?	Part of the requirements will be the situation model described above; in addition, diagnosis information may be required to locate the fault, indicate its cause, explain possible consequences, and provide links to corrective action.	For event location displays: system simulation, causal analysis, and retrieval of advice on consequent and recovery action linked to causation analysis.
How can the failure be contained?	Assuming a dangerous state has arisen, this sub-task specifies how the impact can be mitigated either to stabilize the system or buy more time for decision making.	For dealing with side effects of hazard: evacuating personnel, warning others, shutting down non-essential services.
How can a safe state be recovered?	Recovery sub-tasks are specified to bring the system back to a stable state.	For initiating remedial action: auto-recovery procedures, reporting system status.
Should manual recovery procedures be initiated automatically or displayed for the user's decision?	Requirements issues are how easily and quickly abnormal states can be rectified, linked to the "time to think" problem. If the abnormal state cannot be allowed to persist for more than a few seconds then corrective action must be automatic. This also depends on the extent of the advice that can be provided from the diagnostic phase.	Automated responses, automatic response with over-ride and time-out or display of recovery procedures.

For *monitoring normal system operation*, a system model is necessary because when the unexpected happens the operator needs to be aware of the preceding situation. Situation awareness or keeping the users in the loop is a key concern for automated control systems. The requirements issues are how much detail to provide for the user and how to keep the user engaged with a system that may be mundane and boring for 99 per cent of the time. Generic requirements are for system simulations, interactive system microworlds that can be queried to maintain user engagement. Situation awareness depends on how accurate and comprehensive the model is, but the user does not want to be swamped with unnecessary detail. Requirements issues are update processes that can be customized by operators so that they are not overloaded with data. Answers to the update problem will depend on the complexity and timing of events in the domain.

Using the physical model of the controlled system, questions probe possible *exceptions* to the intended behaviour of the controlled system; see Table 7.5.

Diagnosis requirements involve interpreting the system model with the scenario to understand the reasons for failure. While this will depend on domain knowledge, there are *generic causal factors* that should be investigated in any safety critical system:

- Are the system components closely coupled (interdependent)? Closely coupled systems will have more unpredictable side effects and cascade-type failures. Modular systems with clearly defined interfaces between sub-systems minimize coupling, making failure event tracing easier and failure impacts simpler to contain.

- Are system processes time-critical? Most safety critical systems have real-time processes. Timing needs to be established to decide how controllable the system's behaviour will be from the operator's viewpoint. Events in the sub-second range will have to be dealt with automatically. Events within the minute and second range are still difficult for humans to react to without thorough training.

- Do the system processes have exponential effects? Exponential effects lead to "run away" failure and are often hidden in the early stages of a failure. Clear warnings about possible rapid increases in system parameters should be given before they occur. Simulations can help operators anticipate exponential effects.

- Do the exception events have side effects or cascade effects? Many apparently innocuous events can become dangerous through compounding effects.

This analysis should have discovered requirements for dealing with and preventing hazards; however, the power of the analysis is dependent on the number and appropriateness of the scenarios that are tested. The next section proposes a more detailed analysis that also helps to systematically expand the set of scenarios by suggesting different permutations of rogue events.

7.2.3 Event Pattern Analysis

Requirements analysis needs to deal with exceptions and abnormal event patterns. This section takes the scenario analysis to a more detailed level by questioning the pattern of events. A taxonomy of possible event permutations is proposed that serves to indicate new scenario variations.

Inbound events imply requirements for processes to trap the input and deal with normal and abnormal patterns. The following questions check for possible permutations that may occur and extend the event-driven approach described in Chapter 3. This analysis extends the practice of tracing abnormal paths in use cases, where the simple question is asked "what alternative paths are possible in the use case that deal with exceptions and errors?" (Graham, 1998; Cockburn, 2001). This section unpacks that question to point the analyst towards the many potential types of error and exception that may occur. Analysis starts by investigating Hollnagel's (1993) event phenotypes, and proposes generic requirements that may prevent or remedy the problem. Generic requirements at this level are indeed very general, so they act as pointers towards specification of more detailed requirements; see Table 7.6.

The design implications for these dependencies are well known in the software engineering literature. Dealing with event permutation requires guarded commands. The requirements specification can be elaborated with entity life histories to detect patterns of correct and incorrect events in a sequential order, as described in Jackson's method (1983). As Jackson points out, a filter process can be specified to detect abnormal events by test probes in a normal input sequence; corrective action is taken if an unexpected event is detected. If the dependencies between the scenario and components in the intended system are predictable then validation

Table 7.6 Questions to identify potential errors, with corresponding generic requirements.

Questions	Generic requirements
What happens if the event doesn't arrive? Can the system continue to function? If the event is essential will the system signal a malfunction?	Detection by sequence analysis or time out, signal warning, request resend, assume default
What happens if the event arrives too early or too late? Is the system time sensitive? Can early events be buffered until processed; if so how many can be buffered? Can late arriving events be tolerated? If so can the system halt other tasks and resume on arrival; if not can a malfunction be reported?	Detection by sequence or timing analysis, signal warning, buffer and use (too early), halt and reset (too late)
What happens if an event arrives in the wrong order? Is the system sequence-dependent? If so can mal-ordered events be buffered and sorted into an acceptable order?	Detection by sequence analysis, signal warning, buffer and reorder, halt and reset (strict sequence necessary)
What happens if a duplicate event arrives? Can the system detect duplicates and eliminate unwanted copies?	Detection by sequence or identity analysis, signal warning, eliminate duplicates
What happens if an unexpected event arrives? Can the system deal with unknown input? Can the system interpret extraneous events and report their presence?	Detection by event type/property validation, signal warning, eliminate event, request resend (wrong data)
What happens if a corrupted event arrives? In this case can the system detect that the contents of the message are damaged? Can the system send a request for the event message to be retransmitted?	Detection by event type/property validation, signal warning, delete event, request resend (corrupt data)

requirements for malformed inputs are easily specified; however, if event arrival is random then requirements are more difficult to elaborate. Three main classes of event imply the different responses by the system:

1. Known events which can be validated for the order and timing of their arrival. In this case there is a system requirement to detect the event order against an expected life history and then take error-correcting action.
2. Known events that may arrive in any order. The requirements in this case are to check for the plausibility of events and provide undo facilities to correct user mistakes or warning messages when events appear to be unusual, e.g. an aircraft descends after leaving an airport.
3. Unknown events. The system should continue to function correctly so the requirement is for an exception capture-procedure; or a reporting mechanism so that human operators can investigate exceptional events.

7.3 Modelling Combinations of Influencing Factors

RE for safety critical systems is a laborious process, all the more so because of the volume of scenarios that have to be tested at different levels. This section describes a modelling approach and tool that uses Bayesian belief networks (BBNs) to combine the influencing factors into a more formal and predictive model of human error.

BBNs are graphical networks that represent probabilistic relationships between variables. They offer decision support for probabilistic reasoning in the presence of uncertainty and combine the advantages of an intuitive representation with a sound mathematical basis in Bayesian probability (Pearl, 1988). BBNs are useful for inferring the probabilities of events that have not as yet been observed, on the basis of observations or other evidence that have a causal relationship to the event in question.

A BBN is made up of *nodes* and *arcs*. The nodes represent variables and the arcs represent (usually causal) relationships between variables. The example in Figure 7.5 is a fragment of the net describing the complexity of an action as a variable that is affected causally by two factors: the level of physical detail and the cognitive complexity of the task. Variables with either a finite or an infinite number of states are possible in a BBN, so the choice of measurement scale is left to the analyst's discretion. As demonstrated in the case study (see Section 7.2.1), variables are generally assigned to one of the three possible states: *high, medium* or *low*.

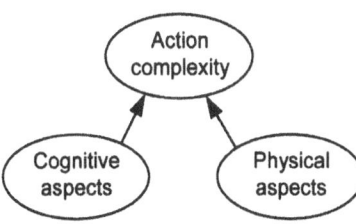

Fig. 7.5 Simple BBN fragment; the two prior nodes have a causal influence on the posterior node of action complexity.

In the illustrated fragment, if we know that both the cognitive and physical aspects of a particular action are high, then the probability of the overall complexity being high is greater than if we know the action has a low level of physical detail and involves little cognitive ability. In the BBN we model this by filling in a node probability table (NPT). Table 7.7 shows the NPT for the *action complexity* node.

Each arc and table of conditional probabilities represents knowledge about one node that is useful for predictions about another node. Hence in Table 7.7 the requirements engineer has asserted that if the cognitive aspect of the action is high and the physical detail of the action is high then the probability of overall action complexity being high is 0.9, medium 0.1, and low zero. The table is configured by estimating the probabilities for the output variables by an exhaustive pairwise combination of the input variables. BBNs can accommodate both probabilities based on subjective judgements (elicited from domain experts) and probabilities based on objective data.

Table 7.7 Node Probability Table (NPT) for the action complexity node.

		Probability of action complexity being		
Level of cognitive ability	Level of physical detail	Low	Medium	High
	High	0.9	0.1	0.0
High	Medium	0.7	0.2	0.1
	Low	0.6	0.3	0.1
	High	0.6	0.3	0.1
Medium	Medium	0.5	0.2	0.3
	Low	0.3	0.3	0.4
	High	0.3	0.3	0.4
Low	Medium	0.2	0.2	0.6
	Low	0.0	0.1	0.9

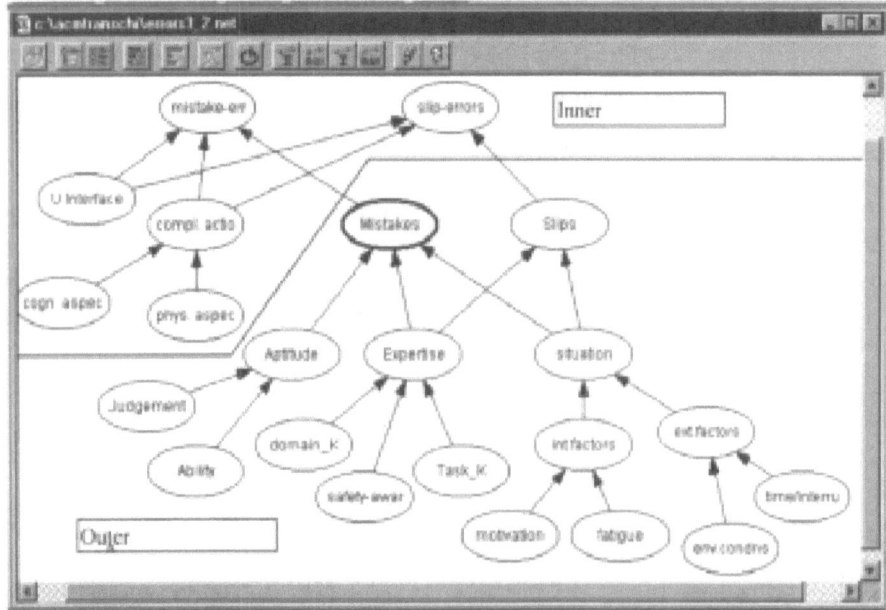

Fig. 7.6 A generic BBN network for human error analysis.

When the net and NPTs have been completed, Bayes' theorem is used to calculate the probability of each state of each node in the net. When evidence is available from particular scenarios, values are entered and automatically propagated through the network, updating the values of other nodes. The taxonomy of influencing factors is used to construct the BBN model either by tailoring the generic model described in this section (e.g. Figure 7.6) or by building a new model from scratch.

The user interface (UI) node represents estimated or measured usability of the UI design; whereas the cognitive and physical complexity describe human actions either with the UI or in a manual task. Physical and cognitive attributes of actions are separated because physical actions require attention and perceptual resources, and hence are more prone to slip-errors (Norman, 1988) in comparison with cognitive actions. Assigning complexity to physical and cognitive actions requires expert judgement. These variables are judged by metric-based techniques, such as cognitive complexity theory (Kieras and Polson, 1985) or by empirical evidence (Bailey, 1982). For example, an action with a high cognitive level would have relatively more decision points and require complex problem solving or judgement. An alternative approach would be to use the Task Load Index of Hart and Staveland (1988). Physical complexity is taken to be a function of the precision, number and diversity of movements required, the degree of sensory motor co-ordination, duration of action, and number of different modalities and limbs involved. Generally, actions that involve multiple limbs and complex sensory motor co-ordination are complex. This complexity judgement interacts with task knowledge in the outer layer, which has to be taken into account when setting up the scenario for a BBN analysis. The outer layer takes a sub-set of the influencing factors (Sutcliffe, 1998) that represents a working hypothesis about the factors that are likely to be more important causes of error – i.e. time pressure and fatigue – following Reason's (1990) work on frequency gambling, with judgement, motivation and knowledge, which are important influences on mistakes and task performance (Bailey, 1982).

In accordance with the views of Reason (1990), Hollnagel (1993) and Rasmussen (1986), the BBN is configured so that the *situation* comprising the relationships between external and internal factors on the operator contributes to slips more strongly than to mistakes. User knowledge or *expertise* is affected by *domain* and *task knowledge* as well as a level of *safety awareness*, and contributes to both types of error but more strongly to mistakes. General *aptitude*, caused by combinations of *ability* and *judgement*, primarily influences mistakes.

The BBN model, once configured, is run against a task script of a particular user's behaviour with the system. In order to do this, each input node is set to one of its alternative states, in this case high, medium or low. For example, of the outer BBN variables, an able user will have the nodes *judgement* and *ability* set to high. Ability is measurable according to the user's level of qualification for the task at hand. Judgement can be determined by a questionnaire inventory such as the need for cognition (NFC) scale (Cacioppo *et al.*, 1984). *Domain knowledge* may be low but *task knowledge* and *safety awareness* will be high in a user familiar with technical and safety procedures in the workplace, but unfamiliar with the technical background. These three measures are obtainable via questionnaires devised at the workplace and specific to the domain and task(s) involved. *Motivation, fatigue* and level of *time/interrupts* being set to high, medium or low has to be an estimate, but preferably based upon historical data of individual operators. *Environmental conditions* was expanded for our case study (see Figure 7.7) as affected by temperature, humidity and dust levels, all of which can be easily measured. The *environmental conditions* node may be set to good, average or bad.

Information may be acquired by a variety of techniques and sources, as outlined in Table 7.8. Questionnaire-like inventories can capture psychological attributes of people, while environmental variables can be taken from past histories or measured directly. Inevitably some variables will have to be estimated if no measures or previous data are available. If the action involves a human–computer interface then its usability is assessed (*U interface* node). Measures for usability can be acquired from evaluation of users' observed problems using methods such as model mismatch

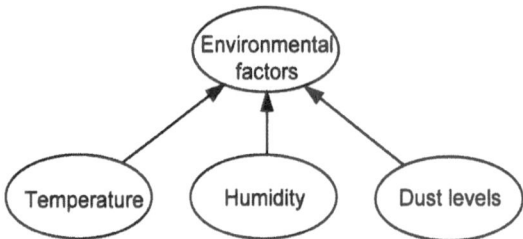

Fig. 7.7 Environmental factors affecting the laser spectrophotometer.

Table 7.8 Inputs and sources of information for each method stage.

Inputs	Source	Method stage	Output
Domain knowledge Safety history	Domain experts Safety documentation	Domain analysis	Key safety critical properties for the domain
Influencing factor taxonomy Generic BBN	Method	Select influencing factors	Causal model as BBN
Generic BBN + NPT tables Prior problems	Method Domain experts	Calibrate BBN	BBN model with configured NPT tables
Questionnaires, experiments, expert judgement	Users Domain subjects Domain experts	Measure/estimate variables	Input values for outer BBN scenarios
Influencing factor variables Scenarios of domain	Estimates or measures Domain experts	Select domain scenarios	1-N scenarios to be run for outer layer BBN
Scenario variables: high/medium/low	See stage 4	Input scenario variables	Outer layer BBN for selected scenario
Interviews, observation of user activities	Users, domain experts, documentation	Task model and complexity estimates/sub-goal	Task steps with cognitive/physical complexity
Sub-goal complexity	Task complexity: measure or estimate	Input sub-goal complexity assessment	Inner layer BBN for selected scenario/task sub-goal
Configured BBN with input variables	See stage 8	Run BBN	Mistake/slip probabilities for task sub-goal
Error probabilities for sub-goal Safety guidelines	See stage 9 Method	Walkthrough sub-goal	UI safety requirements for each interaction step

analysis (MMA: Sutcliffe *et al.*, 2000) or co-operative evaluation (Monk and Wright, 1993). If a prototype of the UI design does not exist, the usability score is set to medium for all potential human computer actions, and low (i.e. no design problems) when actions are unlikely to involve human computer interaction.

Once the outer and inner BBN variables are set, the BBN tool automatically calculates the propagation of probabilities throughout the network.

Note that some of the influencing factors representing the user and task/domain can be set as constants in the application. For example, in our case study in Section 7.2.1, the user is assumed to have little domain knowledge. The value of the *domain knowledge* node was therefore set to low for all test scenario runs. Other aspects of the user were varied, as were all aspects of the environment. The quality of the user interface was set to medium. Mistake-error and slip-error probabilities for each action and for the different test scenarios were then compared.

From the hazard analysis of the laser spectrophotometer system it was found that the system is particularly sensitive to temperature, humidity and dust. The environmental conditions node of the BBN can therefore be expanded as shown in Figure 7.7.

Task analysis of the operator task, *planning the sampling run*, reveals that the operator has to calibrate the machine against a set of standard samples, then go through the cycle of loading samples for each measurement as shown in Figure 7.3. We assume the operator to be of medium ability and aptitude, and having some knowledge of the domain, so the *domain knowledge* node is set to medium. Finally, the quality of the user interface (*U interface* node) is similarly set to medium. Thirty-two test scenarios were then run with different combinations of the following attributes:

- safety awareness (high or low);
- level of task knowledge (high or low);
- internal factors of motivation (high or low) and fatigue (high or low);
- environmental conditions (good or bad) which refer to temperature, humidity and dust levels;
- amount of time/level of interrupts (high or low).

Task action complexity was categorized according to the combination of physical and cognitive aspects of the action as follows:

- *simple* – low physical complexity and low cognitive complexity;
- *physical* – high physical complexity and low cognitive complexity;
- *cognitive* – low physical complexity and high cognitive complexity;
- *complex* – high physical complexity and high cognitive complexity.

The complexity classification of the task sub-goals is given in Table 7.9.

The safety implication for errors are primarily data accuracy for sub-goals 1–4 and 8. Data analysis errors can have safety critical consequences if mistakes are made in analysis, for instance of dangerous chemicals. Sub-goals 5–7 can have safety implications for the operator as the laser light is intense and parts of the body should not be exposed to it. The design therefore needs to prevent laser exposure when the sample cover is open.

Table 7.9 Complexity classification of sub-goals for the system operator task.

Task sub-goal	Cognitive attributes	Physical attribute
1. Analyze measurement	Complex: requires reasoning from domain knowledge	N/A
2. Plan analysis run	Complex: requires planning but some knowledge reuse	Simple: record plan
3. Calibrate machine (also recalibrate)	Complex: but set procedures can simplify	Simple manipulations, including load samples, reading
4. Set parameters	Simple: takes parameters from look-up table based on output from goal 2	Simple: entry of values
5. Load sample	N/A	Simple actions: open cover, place sample on holder, replace cover
6. Take reading (omitted in automatic mode)	Simple: find reading on LCD display	Simple: make note of reading
7. Change sample	N/A	Simple actions: open cover, remove sample
8. Analyze results	Complex: reasoning with domain knowledge to interpret results	N/A

7.3.1 Results of Causal Analysis

The interpretation of the resulting probabilities for errors in the bands "high/medium/low" is calibrated for the domain. In this example, low was taken to be an absolute frequency that would not significantly impede normal system operation; that is, fewer than 0.1 per cent errors, or fewer than 1 error per 1000 actions. Medium was estimated as a range from 0.1–1.0 per cent errors and high was more than 1.0 per cent errors. Figure 7.8 (a)–(e) illustrate the results for 5 of the 32 test scenarios.

Test scenarios 1 and 2 allow a direct comparison of the effects of the external factors, i.e. environmental conditions (heat, humidity and dust) and level of time

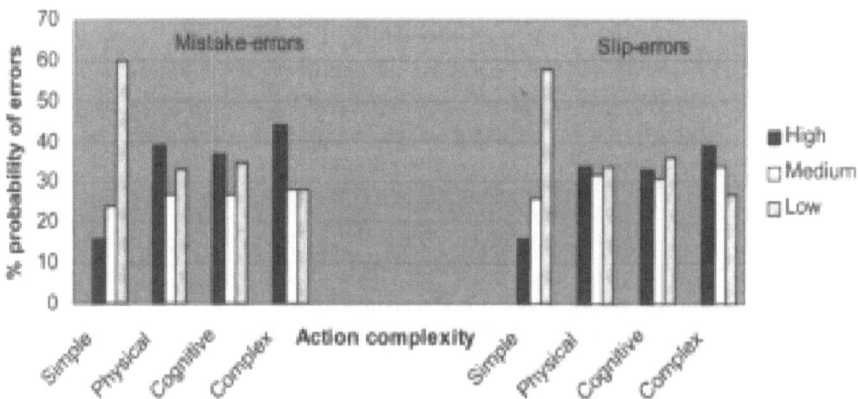

Fig. 7.8(a) Test scenario 1. Best case: operator is safety aware, has good task knowledge, internal factors are good, the environmental conditions are good and the level of time/interrupts is low.

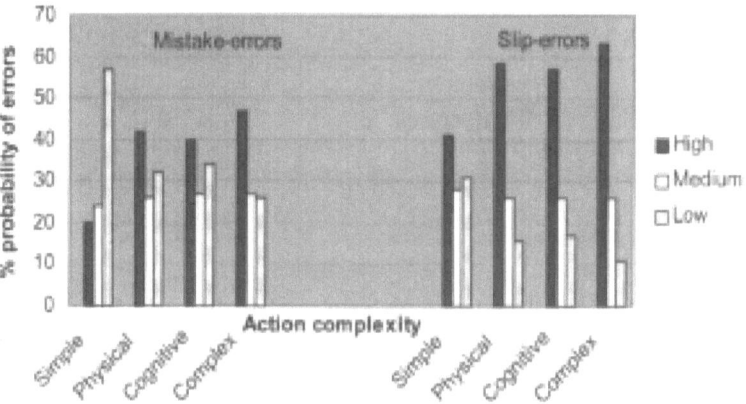

Fig. 7.8(b) Test scenario 2. Motivated operator is safety aware, has good task knowledge, internal factors are good, but the environmental conditions are bad and the level of time/interrupts is high.

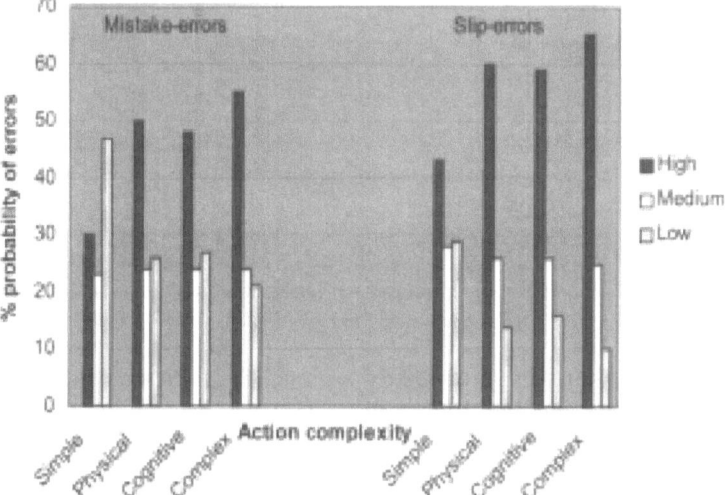

Fig. 7.8(c) Test scenario 3. Well-trained operator with task knowledge is not safety aware. The environmental conditions are good but internal factors such as motivation and fatigue are bad and the level of time/interrupts is high.

constraints/interrupts. In these two scenarios both the operators are safety aware and knowledgeable about the task. They are also motivated and not tired. In scenario 2 however, the environmental conditions are bad and there is little time and plenty of interruptions. The model predicts the probability of both mistake-errors and slip-errors arising in scenario 2, but the increase is much greater for slip-errors than mistake-errors. The probability of more than one slip in 100 operations occurring while performing a complex action-type such as recalibration of the machine, for example, under test condition 1 is 0.39. In test scenario 2 this rises to 0.63.

Test scenarios 4 and 5 similarly allow a comparison of the effects of representing operators with low safety awareness and little task knowledge. They also differ in that the operator in scenario 4 is motivated and not tired whereas this is not the case in scenario 5. This lack of motivation and increased fatigue raises the probability of

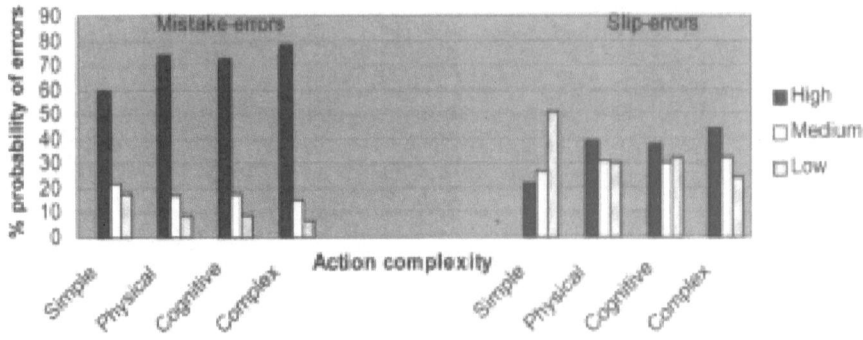

Fig. 7.8(d) Test scenario 4. Operator lacks safety awareness and task knowledge, but the environmental conditions and level of time/interrupts are both good, and the operator is motivated and not tired.

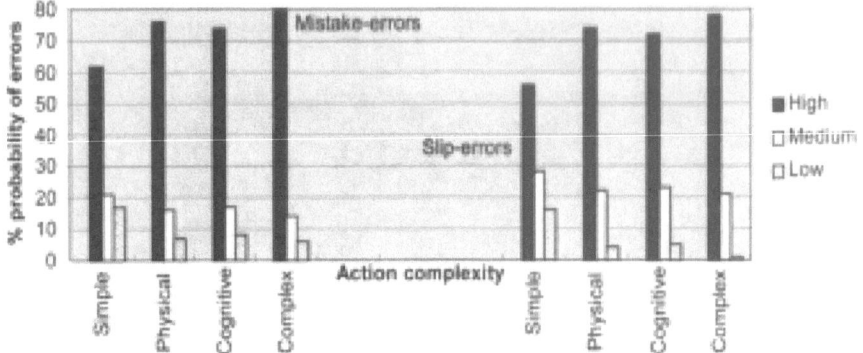

Fig. 7.8(e) Test scenario 5. Worst case: operator is not safety aware, has little task knowledge, internal factors are bad, the environmental conditions are bad and the level of time/interrupts is high.

high levels of slip-errors, approximately 0.1 more than when only external factors were bad. Slip-errors are much lower in scenario 4 and probabilities of high-level range from 0.22 for simple action-types to 0.44 for complex action-types. This implies that the internal factors of motivation and (lack of) tiredness make a substantial difference to slip-errors.

Test scenarios 1 and 4 only differ in terms of levels of operator safety awareness and task knowledge. In scenario 1 the operator is safety aware and has good task knowledge; in scenario 4 the opposite is the case. Otherwise, the environmental conditions and time/interrupts are favourable in both test scenarios and the internal factors of motivation and fatigue are also positive. The results show that the predicted probability of high levels of mistake-errors is much higher in scenario 4 than in scenario 1: from 0.16 for simple action-types in scenario 1 to 0.6 in scenario 4. Similarly the probability of high levels of mistake-errors for complex action-types rises from 0.44 for a complex action-type in scenario 1 to 0.78 in scenario 4. In contrast, slip-errors are less affected. This means that training in both task knowledge and also safety awareness is important in reducing the potential for mistake-errors rather than slip-errors. Actions involving greater cognitive activity, such as analyzing results, are more at risk when task training and safety awareness are reduced.

Test scenario 3 gives predictions of high levels of mistake- and slip-errors in a context where the environmental conditions are good but motivation and fatigue are a problem, as well as there being time constraints and many interruptions. Safety awareness is low, but task knowledge is high. Safety awareness is shown to have slightly more impact upon mistake-errors than on slip-errors.

By examining the results of all 32 test scenarios, it became apparent that time constraints and interruptions had a relatively greater influence in producing high levels of slip-errors than environmental conditions. In general, the implications were, first, demonstrating that adequate training of operators in task knowledge *and* safety awareness is important, and secondly, that good design of work environments and user interfaces enhances motivation, reduces fatigue, encourages sufficient time for each task, and limits the detrimental effects of environmental factors.

7.3.2 Linking Human Errors to Requirements

The BBN model focuses on the probability of human error in user-system interaction. In this section heuristics are proposed to help analysis of the consequences of human error and provide generic requirements to address these problems.

Many problems are hidden and only become apparent when failure in the controlled system (e.g. power plant, aircraft) has occurred. Hence the BBN analysis needs to be linked to a cycle of interaction that accounts for normal operation and human reaction to system failure. To provide such a perspective we have adapted Norman's model of action (1986) to link error types to different phases of action. Heuristics direct the analyst towards the type of error that may occur for different phases in interaction, as summarized in Figure 7.9.

Table 7.10 summarizes the link between interaction stages and generic requirements for both mistake and slip-errors.

The model cycle commences with goal formation and then progresses to "form intention" or a plan of action to achieve the goal. For skilled users these two stages are automatic, so mistake-errors are unlikely. However, even skilled operators may make erroneous plans so the system should try to prevent incorrect or dangerous plans being formed. Slips may occur but they are unlikely to be observable at this stage. The design implications are to provide the control and monitoring functions that the user needs to operate the system. The following challenges are used to assess potential usability problems ranked in terms of impact severity on the user's task; and to assess the normal-mode control system functionality and its user-system interface:

● Does the system provide clear prompts for command actions? (predictability)

● Does the system prevent or make dangerous actions difficult? (protect the user: proactive prevention)

● Is sufficient appropriate information provided for user decisions? (task compatibility)

● Does the operator have to remember routine information? (working memory)

● Is an explicit model of the controlled system provided? (predictability)

● Does the system make the effects of control actions apparent? (observability)

Forming intentions requires knowledge and awareness of the system, so an interactive visual display of the system is advised to counteract low user knowledge. Simulations and decision support models to answer "what if" questions are useful for counteracting mistakes for novice users, and incidentally support learning. The

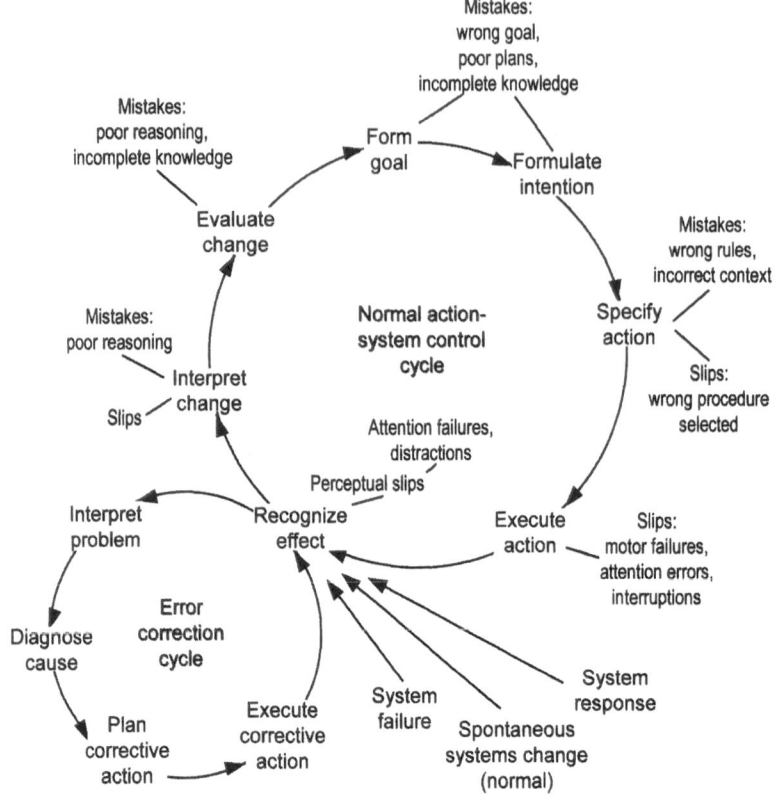

Fig. 7.9 Model of errors pertaining to safety critical interaction.

more information that can be provided about the state of the controlled systems, and facilities to forecast the effect of human action, the better the design will be in dissuading mistaken plans (a real example being the planned but flawed experiment carried out in the Chernobyl nuclear reactor).

Action specification may be prone both to mistakes by less well-trained users and to slips by all operators. Action specification involves deciding the detailed course of action, so complex goals will be especially prone to errors. Well-designed representations of the controlled system help action specification, with transparent detail of the parameters being controlled (e.g. temperature, pressure in a chemical process) and effector devices (e.g. valve, heater, pump, etc.). Mistaken and dangerous actions should be prevented where the effects can be predicted, and warnings built into simulations to forecast the effects of action before execution. Action specification involves faithful presentations of devices and predictable means of action in the human computer interface. This entails clear prompts, cues, commands and operational metaphors for devices. Slips can be counteracted by warnings and lock-outs on undesirable actions (e.g. disabling a heater's "On" switch when the temperature is too high), and by making the effect of an operation immediately apparent and easily reversible.

Action execution is a physical step in the model when the user issues a command or enters data into the computer. The consequence of mistake-errors in the early cognitive stages of the cycle may become visible at this stage. Only at this stage is the user's intent detectable, so prevention of mistakes depends on the ability of the

Table 7.10 The link between interaction stages and generic requirements.

Action cycle stage	Slip counter-measures	Mistake counter-measures
Form goal	N/A	Organization metaphor System model Procedure hints
Formulate intention	N/A	Block unsafe action Aide-memoires System simulations Operational metaphors Clear controls
Specify action	Difficult actions Pre-select safe actions	Prompts cues Device affordances
Execute action	Safety locks Simple manipulations Fail safe Prevent dangerous actions Input validations	Input validations
Recognize effect	Salient warnings Multimodal warnings Clear messages	N/A
Interpret change	Clear messages Repeat warnings	Clear feedback Set change in context
Evaluate change	N/A	System models to reflect change Implications of change Situation awareness models

computer system to predict dangerous effects. Some effects can be anticipated at design time and safety locks or warnings added to the system (e.g. reducing engine power in an aircraft while climbing leads to a stall warning). Potentially dangerous actions should be made difficult. This stage may be prone to slips if the user interface does not give sufficient guidance, or makes the action difficult in terms of sensory-motor co-ordination. Examples of poor design that lead to action execution slips are poor controls, obscure icons, or pointing targets which are too small. The control device should be tested to ensure it is within the limitations of human abilities. Slips at this stage can also be counteracted by constructing devices that make the user aware of the correct actions (see perceived affordances: Norman, 1999), and setting actions in appropriate contexts (e.g. control switch icon is displayed on an appropriate part of a chemical plant diagram). Other error prevention advice is to reduce the complexity of physical action where possible by automated assistance or sub-dividing tasks, and then train operators to acquire the necessary skill.

The next three stages involve user perception and interpretation of system output. Feedback of the effects of user action should be immediately apparent. However, change may arise independently of human action via failure in the controlled system. Slips may occur when the user simply fails to see or hear the warning message. Feedback has to be immediately apparent to the user. Messages should ideally be given in two modalities (i.e. audible and visual warnings). To alert users, the message has to be salient and located where the user will see or hear it. Warning messages have to tread a dividing line between giving messages for true alarms while not giving too many false alarms. This requires careful design of sensors and calibration of monitoring systems.

Once perceived, information has to be understood by the user. In the normal course of the model, the change should confirm that the correct action has taken place, and the effects on the controlled system should be obvious. This follows the observability principle in HCI (Thimbleby, 1990). In the abnormal course at this stage, both slips and mistakes may happen. Slips happen when people jump to conclusions and see what they want to see rather than what has happened. This can lead to capture errors (Norman, 1988) when the wrong course of action is triggered by an interpretation slip. Mistakes involve some misinterpretation when the user's knowledge is insufficient or erroneous. Both types of error can be counteracted by providing clear, simple, but pertinent error messages, using the user's language (Nielsen, 1993). Interpretation is also helped by setting the message in context (e.g. an excess pressure warning is highlighted in a vessel icon on the chemical plant diagram).

Evaluation of change in the normal course involves understanding the effect of action on the controlled system. Forecasting functions may be advisable to project future effects of change to allow the user to evaluate possible impacts on safety, cost, efficient operation, etc. The same advice applies in the abnormal course, but this also implies an embedded diagnosis sub-task. The user has to discover the reason for failure and then take remedial action to return the system to a safe state. The requirements issues in this sub-cycle are summarized in Table 7.11, which is used in conjunction with the following questions:

● Are clear warning and alarms provided? Accurate and timely warnings are vital for any safety system. However, setting the level of warnings needs to be approached with care. Too many warnings will lead to habituation by operators who may ignore the real danger signal or turn the warnings off. Too many warnings triggered by cascade effects will overload the operator with too much information.

● Is problem diagnosis information provided? It is important to provide a clear system model for the operator that displays properties of the controlled system and parameters which are exceptional. The location of hazards within the system should be indicated.

Diagnosis and recovery procedures are checked with the following questions and guidelines:

● Are containment/remediation procedures provided and automatically activated? Operators need time to think when faced with complex system failures that are often unanticipated. Operator training should handle any common expected failures; however, it is the unexpected failures that lead to real problems. Containment procedures should buy the operator time to diagnose the problem. The extent and effect of containment should be clearly communicated to the operator by updating the system model.

● Are recovery procedures or guidance displayed? Although training should make most recovery procedures an automatic skill, display of aide-memoire lists of procedures can help. Human memory may become unreliable when operators are under stress.

● Is the operator given time to take decisions? Human judgement under time pressure is unreliable because of frequency/recency gambling (Reason, 1990), i.e. we tend to use the procedures which are most familiar when under time pressure, even if sometimes they are not the appropriate response. Giving the user time to think is vital for accurate problem diagnosis.

Table 7.11 Error correction cycle stages and system support facilities to prevent mistakes and slips.

Error cycle stage	System support facilities
Recognize feedback	Salient warnings
Interpret problem	Locate hazard, display hazard properties, show affected area, indicate containment
Diagnose cause	Possible causes, display failure history, infer reasons, assumption checks
Plan corrective action	Remedial procedures, warn of side effects
Execute corrective action	Clear corrective controls, highlight appropriate controls, auto-execute corrective action, action checks

● Does the system check recovery actions against the event data? Some actions may be inappropriate in a changed context or for a different system state. If the effect of actions can be validated against the system state and possibly dangerous actions detected then the operator should be warned of the consequences.

Mistakes are especially serious at this stage, as many failures cannot be anticipated and hence the operator has to reason with partial knowledge. Mistakes can be reduced by automated or semi-automated diagnostic tools, which can reason about the causes for error and then recommend, or even automatically execute, corrective action. The degree of automation depends on the designer's confidence in anticipating the probable causes of failure and whether known solutions exist. If confidence in solutions is less than 100 per cent, decision support tools can help users detect possible failure causes and plan remedial treatments, by providing visualization and simulation tools to allow treatments to be assessed before actual execution. Slips in corrective action may also occur when people fail to complete procedures, so semi-automatic checking on remedial procedures and reminders/hints can help.

The cycle of interaction then continues to the next task goal in the normal course of action. In abnormal courses there can be more complexity in terms of containment actions to counteract the effect of system failure.

7.3.3 Case Study: Consequence Analysis

In this section the requirements walkthrough is applied to three different task sub-goals to illustrate how the BBN analysis informs safety critical design.

Plan Analysis Run

This sub-goal has high cognitive complexity, although complexity could be reduced by training with standard procedures. Mistake-errors are therefore likely to be critical when the operators are not trained; predictions of a high probability for medium or high levels of mistake-errors in any scenario therefore need to be viewed with concern. Decomposing this goal with the walkthrough, the first stage – formulate goal – is unlikely to present particular problems, although slip-errors (e.g. forgetting to do this task) may arise from interruptions in the environment.

Forming intentions, or plans for scheduling the run, may be prone to mistakes. The walkthrough model suggests generic requirements for visualizations and aide-memoires of procedures, so the user interface needs to supply checks for planning and examples of typical analysis runs. The selected run needs to be recorded and this requires another separate cycle of human computer interaction to enter the run type into the control system. Slips may occur at this stage, so pre-validation should be used so that the user can select only from available run types in a menu (run types are pre-set according to number of samples, calibration, automatic, semi-automatic, etc.). Mistakes in this sub-goal will have data accuracy consequences later but these are difficult to trap without building an expert system with consider-able domain knowledge to validate the user's plans.

Set Parameters

This sub-goal has low cognitive and physical complexity. Goal formation starts by deciding to set parameters, so output from sub-goal 2 should be available as the nec-essary settings for each run (e.g. sample exposure timings from calibrations). Action specification involves deciding how to find the settings from look-up tables and entering them into the control user interface. Slips at this stage may involve looking in the wrong place for the settings, and number transpositions, or omissions in data entry. Mistakes may arise from misunderstanding the associations between run types and look-up table entries. The walkthrough advises making the cues and prompts clear, and preventing errors by displaying only allowable actions. The valid parameters for the selected run type should be displayed, along with the usual ranges of parameters. Executing the action of entering the parameter values may be prone to slips that can be trapped by making the entered value clear (recognize effect stage), by validation routines to trap unreasonable values (interpret change) or rules to check parameters against each other and known properties of the run (evaluate change). Simple simulation/forecasting tools may also prevent errors, such as dis-plays that link the entered parameters to a selection of successful previous run types to allow the user to cross-check their understanding. Slips in the "recognize effect" to "evaluate change" stages imply the need for clear and legible displays.

System malfunction is possible if the data entry mechanism encounters a software error or the laser control software malfunctions. Detecting alternative paths implies the need for a monitor system to assess the operating state of the laser control software, and signal warnings to the user. To prevent interpretation mistakes the problem should be set in context; for example, a diagram of the system is shown to highlight the fault in the laser control software with simple error messages, possible diagnostic causes and repair suggestions.

Replace Sample

This sub-goal was classified as simple physical action. Mistakes do not apply; however, action slips can have serious consequences if the design does not prevent them. Once a sample run is initiated the first two stages of this sub-goal are simple: an intention to replace the sample followed by the action specification of removing the cover and loading the sample. Safety in the execute action stage can be ensured by design guidelines for making dangerous actions difficult, so design of the physical casing makes it hard to place one's hand, let alone eye, in the laser beam; in addition a hardware/software lock prevents laser beam activation once the cover is opened. Another possibility is to lock the cover while the beam is active. The remaining part of the normal course raises few implications beyond the fact that the effect of removing the sample should be clear to the user. Note that this may conflict with the safety requirement to prevent dangerous actions that suggests the sample container area should be difficult to access and not easily visible. Design frequently involves trade-offs between safety and other non-functional requirements.

More interesting is the possible abnormal course walkthrough at this stage. Taking the design of software laser lock that was activated by the cover being opened, what would happen if this were to fail? The method suggests making the consequences of failure clear to the user through recognizable, interpretable messages or not allowing the failure to occur, through defence in depth. This suggests that a status display should show which component has failed, with an audible/visual warning when failure is detected. Defence in depth could be implemented by an infra-red beam behind the cover so, should the safety lock fail, when the operator's hand intersects the beam a second safety lock is activated to turn off laser emission.

This concludes the case study, which illustrated how the BBN analysis first highlights particular scenarios where errors may occur due to certain combinations of operators and the task environment; then, given a set of tasks, how the BBN predictions of mistake- or slip-errors can be interpreted in a walkthrough of the user's task to consider which safety critical guidelines might be recruited to the design. While this approach does not claim to be comprehensive, as only a selection of available safety guidelines have been included (Leveson, 1995; Hollnagel 1993), it does provide a framework on which more comprehensive and targeted advice can be delivered.

7.4 Formal Reasoning about Safety

No chapter on dependable systems would be complete without mentioning the role of formal methods for checking and refining safety critical requirements. Although the approach in this chapter has been primarily informal, formal methods really come into their own for safety requirements. A detailed model of the system has to be created, so that model checkers can be run to test conditions and states that must happen (liveness conditions) or to check that dangerous unsafe states do not occur.

Many formal languages have been applied to requirements analysis for safety (see Dubois *et al.*, 1997; Heitmeyer *et al.*, 1996; Leveson, 1995); I refer the reader to proceedings of Requirements Engineering and Safety Critical conferences for copious examples. Although it is not my intention to review this area, interesting applications of formal approaches include, first, the use of *counter factual reasoning* (Johnson, 1999), which examines arguments in system models, and then infers the consequences if the argument is negated. In other words, it poses the question, "What would have happened if the argument did not apply, or this fact were not true?". As we don't naturally reason well with negatives (remember confirmation bias in Section 2.5), counter factual reasoning can be a powerful tool in analyzing flaws in system defences. Second, *deontic logics* express the obligations of agents to behave in a certain way, so these also provide an important means of checking the dependencies between components and the conditions under which an agent should exhibit a particular behaviour (Van Lamsweerde and Letier, 2000). Finally, many safety critical problems are also time sensitive, so *temporal logics* are useful for reasoning about time intervals and periods during which the system must respond to events.

The problem with formal reasoning is the effort expended in creating the formal model in the first place and the complexity that can be dealt with by such processes. Formal models of deterministic software systems are amenable to model checkers and proof systems; however, systems usually involve people. Creating a deterministic model of human behaviour exceeds the current level of knowledge in psychology, so we have to settle for more approximate probabilistic models. Hence a combination of Bayesian models for the social system, as described in this chapter, with formal models for the software systems may be the way forward. Integrating informal and formal models helps bridge the explanation gap, and this has proven successful in safety critical reasoning with logics and design rationale (Johnson, 1996). In some cases, however, the complexity of software defeats formal modelling. In these cases the requirements engineer may need to address the problem via architecture and specify requirements for safety kernels, which are critical modules implementing key aspects of safe behaviour. Safety kernels can then be subject to the rigours of formal proof and testing.

7.5 Summary

This chapter has investigated requirements for safety critical systems, which elaborate the usual concept of functional requirements by adding requirements for error prevention and defences against failure. As safety is a key non-functional requirement in many systems, requirements engineers need to pay particular attention to the specification. A method for safety critical requirements analysis was presented and illustrated with a case study of defining requirements for control of a scientific analysis instrument. A taxonomy of influencing factors was described that can be used to reason about the causes of failure and human error. Most accidents have several causes that provide a window of opportunity for a rogue event to occur. Requirements for dependable systems need to anticipate these causes in managerial social issues, human characteristics and the system environment, as well as properties of the design. The influencing factors at each level in the analysis were cross-referenced to generic requirements to prevent, or at least counteract, the potential cause of failure.

Analysis by influencing factors can be improved by creating predictive models from the taxonomy. A BBN model was described that predicts the probability of human errors based on scenarios of influencing factors in the system environment, a profile of the human operator, the task and properties of the design. Output from the BBN model can be used in a walkthrough analysis of user-system interaction to plan defences at each step in an operational sequence. This analysis uses question prompts at each step that are linked to generic requirements and design guidelines taken from the dependable system design literature. Scenarios are a useful means of testing system defences and validating requirements in safety critical systems; however, this raises the problem of defining or acquiring a sufficient set of scenarios. The influencing factor taxonomy can provide one means of reasoning about the coverage of scenarios.

8 Future Directions

In this final chapter I will speculate on the future courses that requirements engineering may take. This is not a matter of imagining the impact that future technology may have on RE, however; rather, it is motivated by revisiting the fundamental problems of RE: communicating, understanding and transforming needs into designs. RE will face more pressing problems as systems become more complex, distributed and ubiquitous. It will become increasingly difficult for any one person to understand such complexity and then specify requirements for it. We therefore have a dilemma of what I call the *horizon of knowability*: as technology advances so does our ability to design more complex systems, yet our ability to understand such systems is finite. We are limited by our cognitive capacity and the time taken to understand complex systems; this capacity can only change slowly as we devise better methods. In contrast, the rate of technological change is exponential, as it feeds progressively on its own baseline. We tend to acknowledge this as the accelerating pace of change in society. So where does the answer lie? I will argue that we should look in two possible directions: first toward *artificial intelligence and learning machines*, and secondly towards *automated design environments* that communicate with us in natural language and graphics. Both have a common implication: in the future, requirements engineering will become progressively concerned with the design environment, while what we currently consider to be requirements analysis will either be learned automatically or achieved by conversation with an application generator machine.

8.1 Requirements and Design

Requirements engineering has tended to regard itself as the beginning of the design story. In other domains such as product design, however, this is rarely the case. Design often starts with market surveys and concept definition. Although enterprise modelling has attempted to close the gap between business-oriented requirements analysis and RE (Loucopoulos and Karakostas, 1995), these methods still have their intellectual roots in conceptual modelling rather than in business theory and marketing analysis. While RE can be connected to business modelling and socio-economic theories, this has only been attempted for defining high-level requirements for CSCW systems that support inter-organizational relationships (Sutcliffe and Li, 2000). In the future, marketing theories and models will need to converge with RE methods so that early product definitions are driven by business-oriented motivations. Some moves in this direction can be seen in frameworks for product concepts and market maturity models, for instance Norman's (1999) proposal for information appliances. More detailed suggestions have started to appear in e-commerce, where management scientists have investigated how information presentation requirements can be driven from market models (Lohse, 2000). As success in e-commerce becomes increasingly driven by business values of trust, corporate image and brand identity, RE methods will have to adopt business influences.

Requirements analysis has tended to focus on functionality of transaction processing or real-time systems but many systems involve information and supporting users' decision making. Decision support models and requirements analysis for such systems (Nunamaker *et al.*, 1991) also need to be added to the battery of RE techniques. The role of information is not prominent in RE, yet in many systems information provision is the prime requirement. RE will need to incorporate knowledge from information scientists (Marchionini, 1995) and adapt models of information analysis (Sutcliffe, 1997, 2000). As the sophistication of software advances, decision support systems are becoming increasingly intelligent with embedded expert systems (Singh and Zeng, 1996). When the intelligence is hidden from the user, requirements analysis converges with knowledge acquisition. Convergence with knowledge acquisition methods (Breuker and Van Der Velde, 1994) has been slow, although some techniques have found their way into RE (Maiden and Rugg, 1994). When the decision lies in human hands, however, we need better models of trust and discourse to decide how proposals from computers should be communicated to people so they co-operate effectively.

E-commerce and the web have already imposed new demands on RE. Functional requirements will take second place in the future to information requirements and new types of non-functional requirement. A key problem will be attracting users and then persuading them to take a particular decision, make a purchase, etc. Persuasive computing and captology is already a growth area of research and commercial concern (Fogg, 1998; Tseng and Fogg, 1999). Requirements engineers will have to understand the customer in greater detail than is currently achieved in marketing, and then specify how the design might motivate users. This will inevitably propel RE towards user interface concerns and design for attractive, aesthetically pleasing and motivating systems. Research has demonstrated that people treat advanced user interfaces employing speech and human images as a virtual human rather than a computer (Reeves and Nass, 1996). The role of human personality and how it is represented in image, speech and animation is becoming a new area of design that can be deployed to persuade and influence, extending the intuitions already practised in television commercials.

RE will therefore need a new focus on building models of the user population and specifying requirements in light of their motivation. Psychological theories of motivation will be necessary to specify which design features will attract and persuade users. Such design features will include ideas on aesthetic design to attract users, and an understanding of the role of brand identity and corporate image. Requirement engineers will have to work closely with designers to refine requirements via storyboards and early prototypes. Theories of motivation (Maslow *et al.*, 1987; Orth, 1997) can give guidance about how different user needs and motivations can compete for the user's attention (Sutcliffe, 2001); however, these have yet to be exploited to create methods and techniques for RE for persuasive computing.

8.2 Requirements for Service-Oriented Software

The world of business, indeed the world in general, is changing at an increasingly faster rate. Businesses struggle to bring products to market, and we all struggle to deal with new technology and ways of working. Software has to keep pace. But we find software legacy systems hard to change, and some software systems that take years to develop become obsolete before they have been implemented. Software systems in the future will have to evolve. There are two main responses to this problem: either build generator systems which evolve by creating new versions from high-level descriptions of requirements; or develop component-based engineering where new versions can be designed by flexible assembly of reusable components. The trade-off is between component-based engineering in which development is achieved by selecting components from a reuse library and composing applications, and application generation that envisages a high-level requirements language that is automatically transformed in executable code. The former implies that requirements analysis will converge with information retrieval as a matching process of finding appropriate components for a new application, while the latter places more emphasis on an approximation of natural language to express requirements, while the code itself may be a consumable that is disposed of when a new version is necessary. Software evolution poses an interesting conjecture about the economics of different design approaches that I have addressed in more depth elsewhere (Sutcliffe, 2002).

Time to market is an ever-pressing mantra for software developers who have less time to think and clarify requirements as well as less time to construct well-designed software. The result is poorly designed software that no one wants and everyone has to struggle with. Given the time pressure on developers, software reuse rather than re-inventing the wheel seems to be an obvious solution. Unfortunately, this has not been the success story many had expected (Tracz, 1995). RE has hardly acknowledged requirements for reusable components, yet in the future the need to specify components that are reusable may become an essential part of the design process.

One view of the future conceives software as a service (Brereton *et al.*, 1999). Software may become a set of downloadable services that can be composed to create applications. Software is almost a consumable in the marketing definition: you download it, presumably pay for it, and then throw it away. But what happens when we unpack this vision? First, it begs the question of what a service is. Ideally, software should act as a well-trained servant who understands one's requirements perfectly and then delivers the required goods or request efficiently but unobtrusively. So at the core of the concept is understanding requirements. There is the assumed intelligence of finding the appropriate solution for the request and delivering the service in an efficient manner.

The aim for all design is to achieve an optimal fit between the product and the requirements of the customer population. Leaving aside the problems of market development and customer education, generally the better the fit between users' needs and application functionality, the greater the users' satisfaction. Product fit will be a function of the generality/specialization dimension of an application. This can be summarized in the law of user satisfaction:

The user satisfaction supplied by a general application will be in inverse proportion to its complexity and variability in the user population.

Unfortunately, many applications, and especially Microsoft products, suffer from functional bloat because the requirements process is driven by bug fixes and modification requests from a minority of power users, rather than by any systematic analysis of what the majority of people want. Applications also suffer from their legacy. Even if Microsoft wanted to simplify Office products, as they have with cut-down word-processors, most people still identify with the product that is most familiar, even if it is worse than the competitors'. One escape is to make software a composable service instead of an off-the-shelf product. The ideal is to have applications composed to fit our individual needs; in contrast, general products become progressively less satisfactory as they try to cater for more and more users with diverse needs. The consequences flowing from this law are that design by reuse will be more difficult with a heterogeneous user population, because getting the right fit for each sub-group of individuals becomes progressively more challenging and expensive. The second consequence is that larger-scale and hence more complex applications will be more difficult to tailor to users; also, complex products will impose a larger learning burden. The implications are that complex general products tend to make us frustrated because we cannot easily get them to do what we want. Furthermore, the more different types of user there are in a population (gender, age, different cultures), the harder it is for a general product to suit all. The reaction of designers is to strive for adaptable products, but that just increases complexity either overtly by giving users extra work filling in extensive profiles, or covertly by intelligent adaptive interfaces that rarely guess exactly what we want.

The second law relates to the effort users will be willing to devote to improving the fit between their requirements and the product:

The effort a user will devote to customizing a software product is inversely proportional to its complexity and the variability in the user population.

This implication follows on from the first law, in that the more complex a product is the more effort we have to devote to customizing palettes, user profiles, setting parameters, etc. Furthermore, general products may not motivate us to spend much time on customization because the utility they deliver is less than a perceived ideal of our specific requirements.

We can formulate this problem as a third law:

User effort in customizing and learning software is proportional to the perceived utility of a product in achieving a job of work or entertainment.

This implies that part of the requirements problem is not only to specify functionality but also to explain it in an attractive manner so people want to use the function. People will devote considerable effort to learning how to use a product even if it is poorly designed, as long as they are motivated. Motivation itself is a complex subject. Classical models (Maslow *et al.*, 1987) point towards several layers of motivation directed towards different rewards, ranging from basic needs (hunger, sex) to higher order self-esteem and altruism. Teenagers already spend considerable time learning and adapting games software that rewards them with excitement; however, most business software only rewards us by getting a job done. If the effort expended in customizing software is not motivated by a direct reward, we tend to resent this imposition on our time. Our motivation will depend critically on perceived utility and then the actual utility pay-off. For work-related applications we are likely to spend time customizing and configuring software only if we are confident that it will empower our work, save time on the job and raise productivity. Some adaptation/customization effort may be off-loaded on to the customers, but only when the perceived utility (i.e. the customers' expectation for the product before they have used it) is high. Unfortunately we rarely have any confidence that customizing a product will help. For personal productivity tools, the motivation may be higher than for more general-purpose business software that is regarded as part of the job. Technology acceptance models (Davis, 1993) point out that decisions to adopt technology depend on utility, usability, and possibly fun. Given that fun is going to be limited in many work domains, utility and ease of use will be critical for competitive advantage. This will add a new dimension to non-functional requirements, since many applications in the future will not only have to be usable but will also need to be attractive. Requirements for aesthetic design, visual style and image will need to be accounted for.

Adaptation causes users work. Software should therefore spare us the burden. Intelligent agents can automatically detect our requirements and provide appropriate services. Unfortunately adaptation is a two-edged sword. It is fine so long as the adaptation is accurate and fits with our requirements, but when the machine makes mistakes and adapts in the wrong direction it inflicts a double penalty. First we have to understand what has happened, as the new dialogue, feature or function is unexpected. Then we have to adapt our behaviour to work around the system's misguided initiative. If this were not bad enough, we also feel resentment at the "damned machine" taking initiative and getting it wrong. This soon gives an adverse image of the computer and lowers our motivation to use it effectively. Anyone who has experienced the Microsoft paper clip "Office assistant" will know what I mean. This leads to the fourth law:

> *The acceptability of adaptation is inversely proportional to system errors in adaptation, with the corollary that inappropriate adaptation inflicts a triple penalty on our motivation (cost of diagnosing mistakes, cost of working around, and negative emotions about systems usurping human roles).*

Hence adaptation cannot afford to be wrong, but adaptation is one of the most difficult problems for machines to model. Even if the system could gather perfect knowledge of its environment and had perfect interpretation, other (human) agents may change their minds about their requirements. Such change cannot be anticipated short of clairvoyance. Some system mistakes are therefore inevitable in adaptation. Given the heavy penalty for mal-adaptation, this approach to software evolution may be ill-advised unless there is a high user motivation that will tolerate mistakes. An interesting aside on the adaptation problem concerns safety critical systems. In dangerous or time critical systems, software has to adapt to events in the environment and take decisions when humans are too slow. Adaptation is taken on trust in many cases but, when the consequences of machine mistakes become apparent in unusual circumstances, the results can be catastrophic (Reason, 1997). This has led many to distrust complex command-and-control systems for avionics, nuclear power plants, and so on.

Reuse, adaptation and customization will all have implications for RE by widening the scope of requirements beyond a set of fixed function requirements towards scenarios for future adaptation and reuse. Indeed, envisioning future uses in scenarios and concept demonstrators may be one way to address this problem. Whatever the concept of services and requirements we need, the main future problem for

RE will still be expression. If only we could express our requirements in the first place then maybe the application could be composed to fit our needs without having to be adapted or customized.

8.3 High-Level Requirements Languages

The ideal in this case is to have a natural language understanding system linked to an application generator. Unfortunately this implies considerable artificial intelligence and currently intractable problems in general-purpose natural language understanding, as demonstrated by the CYC project (Lenat, 1995) that has devoted many years' research to natural language understanding and common sense reasoning, without much success. The trade-off is between the constraints imposed on users in restricting their means of expressing their needs, and the accuracy of requirements statements for machine interpretation. At one end are formal requirements languages that have been developed for specific domains (Freeman, 1987; Liete, St Anna and De Freitas, 1994; Neighbors, 1984), or more general-purpose expressions in Taxis/RML (requirements modelling language: Greenspan, Borgida and Mylopoulos, 1986). These impose a learning burden on users who have to translate their requirements into an essentially foreign language. In some cases a domain language has been used to record a domain analysis to specify the rules, components and constraints to generate applications in particular areas such as air traffic control scheduling and flight planning (Smith, 1992). An example of these complex requirements languages is the plan calculus in the Requirements Apprentice (Reubenstein and Waters, 1989) that was used to specify detailed requirements for library systems but not in a form that users could comprehend.

8.3.1 Natural Language Requirements

Generation necessitates a soundly based means of expressing requirements; see Figure 8.1. This has been researched for many years, and domain-oriented languages have been developed for specifying electrical circuits and formatting text documents; however, the scope of domain-oriented languages has not dramatically increased over the years. Unfortunately domain-oriented languages do not really express high-level requirements; furthermore, they require considerable knowledge of the language and its syntax to specify a design. However, specialist requirements languages do prosper in expert domains where the language is already part of the users' knowledge, for example circuit and hardware design (Anderson and Tourlas, 1998).

The question for designers of future generator environments is which requirements language to choose. Sub-languages based on natural language will be easier for users to understand but these languages are restricted to narrow professional domains. Specialized technical languages will be more acceptable for users where the requirements language is already part of the users' knowledge. However, this still leaves a large class of applications for which no specialist language exists. For these, general-purpose requirements languages or end-user programming languages may be the answer.

Speaking to a machine and having it automatically deliver a faultless application is the holy grail of RE. General-purpose natural language understanding will be

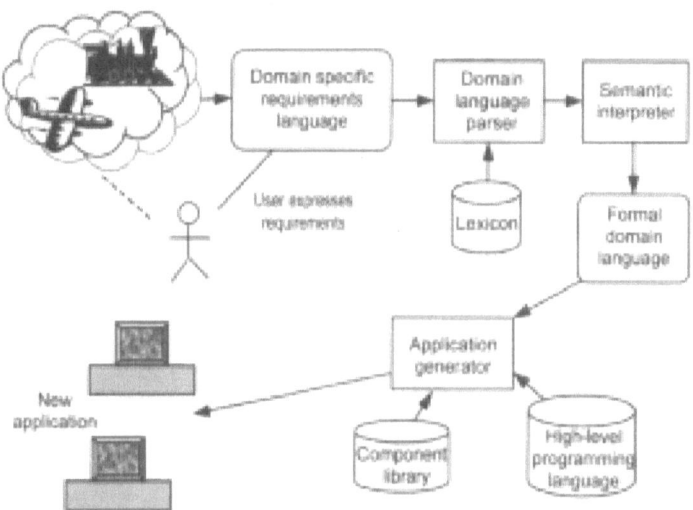

Fig. 8.1 The relationship between application domains, requirements languages and generative architectures.

limited for several years to come by the semantic/pragmatic interpretation problem. Understanding a requirements statement such as "pro forma invoicing will be enforced for all new customers" requires considerable knowledge about what "new customers" means, to say nothing of "pro forma invoices". While this is likely to be impossible for many years to come, it is useful to consider the barriers to such progress. The principal barrier is the AI problem of general-purpose learning machines: a learning system is only as good as the knowledge it already possesses (its domain knowledge). This leads to the knowledge acquisition bottleneck, and attempts to address this problem have met with limited success, for example the CYC project (Lenat, 1995). However, even if we could build a general learning machine, requirements have to be translated into designs in the real (human) world by a process of dialogue and negotiation. An automated requirements assistant would therefore have to converse with users and show them prototypes, while gathering criticisms and suggestions for improvements before redesigning the system. Even if advanced DODEs (domain-oriented design environments) could be equipped with sophisticated natural language processing, there are multiple stakeholders in many applications. Requirements have to be negotiated, prioritized and agreed in a sensitive manner that involves dealing with politics and managing social situations. Even human requirements engineers can experience limitations when facing such challenges, so if we cannot always succeed ourselves, the chances of instructing a machine how to achieve agreed goals is slim. Much of the requirements gathering, negotiating and validation process will remain in human hands.

Inferring data models and class structure from natural language is achievable (Black *et al.*, 1987); unfortunately, specifying behaviour is not so easy. The multitude of ambiguities that English and other natural languages offer make for inaccurate specifications. The way forward may be to use simulation, and graphical representations to explain behaviour of operations on objects to the user. This, combined with explanation on the semantics for specifying behaviour, may help to bridge the understanding gap.

More specialized, domain specific "sub-languages" may hold the answer. Sub-languages are sub-sets of natural language that have limited vocabularies, restricted syntax, and telegraphic expressions to code specific meanings (Grishman and Kittredge, 1986; Harris, 1989) so the user does not have to learn a foreign vocabulary or syntax. An example is the language of air traffic control that is terse and ritualized. Instructions are easily understood by the professional in the domain, e.g. "BA 3450 turn right heading two five oh, descend 10,000" will be understood by the pilot of the British Airways flight as an instruction to turn right on to compass bearing 250 (south west) and descend from the current height to 10,000 feet, but no further. Sub-languages are familiar to professionals and sufficiently terse for limited natural language processing. The narrow domain enables an expert system to be built so user statements can be interpreted accurately. Indeed many commercial expert systems exist in sub-language domains.

However, there is a danger in sub-languages. They usually express operational concepts in a domain, i.e. command instructions, and reports, rather than design requirements. So considerable human expertise is still going to be necessary to interpret the design implications implicit within sub-language utterances. Furthermore, the domain restriction is the Achilles' heel of the approach. Many system generators have to be developed, each targeted on a specific domain. Strangely, this approach has seen little research in software engineering. Generative technologies have only been successful in design domains where the users will accept a specialist requirements language and domain knowledge is well known. Some examples are high-level hardware definition languages that allow automatic creation of electronic circuits and chips (Edwards, 1995), and transformation of database schema definitions into database implementations (Batory, 1998). These specialized generator applications have not enjoyed commercial success compared to general-purpose application generators which have grown out of the fourth-generation language movement (e.g. Powerbuilder) and automatic transformations of UML specifications into executable code (Kennedy-Carter, 2000). Although these products impose considerable burdens on users who have to learn a new (programming) language, end-user computing empowered by such languages is one future approach.

For technical domains, application generation systems will be worth the investment if the domains are reasonably stable over time, for instance in civil, mechanical and other branches of engineering. In other domains where sub-languages exist, generative architectures may also be worth the investment if reasonable stability can be assumed. This seems to be an opportunity for software engineering research, as many of these domains currently develop systems by informal software evolution over versions, for example air traffic control systems and many other command-and-control systems.

In summary, requirements engineering by application generators is but one of the competing approaches for the future. Application generation from requirement languages imposes a high initial design cost for building the generator, but low design costs for each application. The user task fit depends on expressibility of the requirements language, although there are bound to be domain limitations.

8.4 Multiple Methods

RE has already evolved a range of methods and techniques from the formal to the informal. Many have argued for a multiplicity of approaches to deal with the differing needs for requirements as they progress from vague initial ideas to well understood and detailed specifications (Fickas and Feather 1995; Pohl 1993). In the near-term future, RE will probably evolve along its current direction as a toolbox discipline that provides a variety of techniques, methods and tools to address requirements in different types of application.

The informal approach to capturing requirements has made considerable progress as understanding has increased about how prototypes, storyboards and scenario-based approaches can be used more effectively (Sutcliffe, 1997; Carroll, 2000). The dominance of object-oriented design and UML as the *de facto* norm for expressing specifications graphically will continue to focus RE techniques around use cases and scenarios. Executable UML (Kennedy-Carter, 2000) can improve the efficiency of the development process, although executable specification tools do not avoid the pain of expressing algorithmic detail in a programming style language. The natural join would seem to be between rigorous formal specification languages (e.g. Z: Diller, 1994) and informal, but more tractable diagram languages. This course has been tried, but without much success (Gaskell and Phillips, 1995), probably because the research did not harness the power of the formal requirements language to drive a natural language dialogue to clarify facts with the user. Formal specification inevitably raises questions of interpretation that cannot be answered in an inadequate, semi-formal diagram. Formal languages for requirements specification have adopted front-end representations that make them more tractable to domain experts; for instance, table-driven formats are a reasonably intuitive way to express formal relationships and constraints (Heitmeyer, Jeffords and Labaw, 1996). Coupling formal requirements specifications with informal animation is another approach that has shown some promise (Dubois, Dubois and Zeippen, 1997); however, whereas animation of event sequences makes behaviour of objects and agents visible for user validation, making states inspectable is more difficult. States and hidden modes are not easily represented, yet they are often vital for interpreting observed behaviour. Coupling formal and natural language would seem to be the answer, but this has not been attempted, I suspect because formal methods and natural language researchers come from very different areas of academic endeavour.

In spite of the unsuccessful union of informal and formal approaches, there is no reason why these approaches should not co-exist. Indeed, a diversity of methods and techniques is necessary to deal with the range of requirements problems in the real world. Formal approaches will be necessary for safety critical systems where proof of required behaviour can save lives (Easterbrook and Callahan, 1997); in contrast, resource intensive formal specification will not be appropriate or necessary for business systems where some unreliability may be tolerated.

8.5 End-User Development

If requirements expressed in natural language are inherently ambiguous and difficult to interpret, an alternative approach is to provide users with a graphical environment that allows users to specify their requirements by demonstration. This idea has been followed in the programming-by-example (PBE) community. Requirements for end-user development (EUD) can be divided into two phases. First, there is requirements analysis for building the EUD environment; secondly, requirements are either supplied by the user using the EUD environment or captured indirectly from the user by a variety of lightweight learning approaches (see Lieberman, 2001). One approach requests users to demonstrate the before and after state and the PBE system infers the transformation which is then stored as a reusable program. Another approach is to track the user's actions and recognize regular patterns by statistical techniques (e.g. Markov chains, Bayesian nets). This approach has been a favourite for web browser tracking systems that try to guess the user's requirements from typical usage patterns.

PBE/demonstration-based programming is a form of weak learning, but as with many learning systems the problem lies in inferring semantic knowledge from syntactic observations, i.e. a set of movements on a graphical user interface means copy the file from folder A to folder B. This form of learning can either be undirected when the computer records all user actions and tries to infer the more common or significant ones, usually via a statistical approach; alternatively, learning can be directed by the user informing the system when to record significant actions. Undirected learning systems suffer from the false guess problem, because many patterns detected by the system may have no significance for the user. The Microsoft help system uses BBN (Bayesian belief nets) as a predictive tool, yet most users find its "helpful" suggestions unhelpful.

In contrast, more design-oriented approaches provide users with a set of building blocks and a means of composing these into meaningful applications. Domain-oriented design environments, pioneered by Fischer and his colleagues, are a good example (Fischer et al., 1992). DODEs consist of a design work area, a set of design templates, critics that embed design knowledge and advise the user when mistakes are detected, and tutors that guide users through the design process. Applications have ranged from kitchen design to network and user interface design (see Figure 8.2).

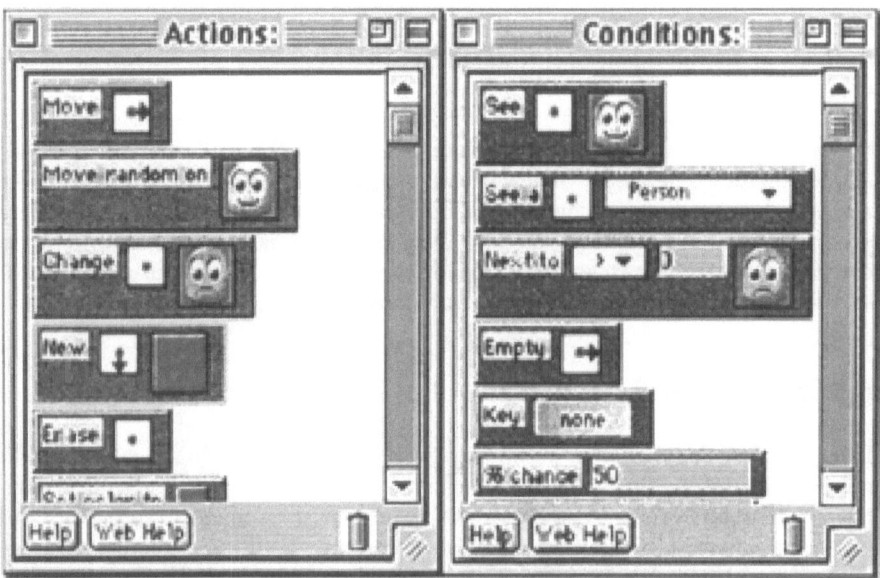

Fig. 8.2 DODE for end-user development.

Agentsheets (Repenning, 1993) provide a graphical scratchpad so the user can visually compose a system, such as a train control simulation by picking train, track and other icons from a palette and then demonstrating rules to the system about points as switches, signals for controlling trains on tracks, and so on. The system can then infer analogous transfer of rules to a similar domain such as a car-driving simulation with cars, roads, junctions and traffic lights. However, the weakness of graphical end-user development environments is the need for a comprehensible spatial domain. This approach might work for designing a logistics system that has a spatial model, but more abstract problems, such as ticket booking/reservation systems, present further difficulties. The former can be expressed as constraint-based problem-solving algorithms with spatial maps, but expression of abstract concepts in maps and diagrams is not immediately comprehensible for novices (Sutcliffe and Maiden, 1992).

Requirements for end-user development involve domain analysis and gathering requirements for a reusable set of components. The requirements engineer also has to specify design support tools and anticipate the set of functions a designer/user may want. This raises difficult problems of scoping domains and deciding the level of abstraction for reuse, for which little guidance exists. I have addressed these problems elsewhere (Sutcliffe, 2002), so here I will just note that requirements for end-user development can work in well-bounded domains when people are likely to agree where the boundaries lie.

Requirements for end-user development assume that a stable domain can be analysed *a priori* and that an adequate set of design knowledge can be gathered to advise the user/designer. While this may be possible in many domains, it still implies that the user/designer will be constrained by the original analyst's view of the domain. The user/designer may find this to be an unnecessary impediment in many cases so this may prove to be the weakness of end-user development. Also these tools are complex and expensive to build; however, more general-purpose development tools such as application generators and fourth-generation languages have enjoyed considerable success (e.g. Powerbuilder, and web development tools ColdFusion, FrontPage, DreamWeaver, etc.).

8.6 Comparison of Approaches

The future approach to software evolution comes down to a few simple choices. On the one hand there is the language process approach in which users have to express requirements and automated processes do the rest. On the other hand there is the structural reuse approach: give users components and let them build applications by themselves. Both approaches impose some learning burden on users and designers. These costs are summarized in Table 8.1.

If designers follow the component engineering route then they have to learn about the properties and facilities offered by components. Self-explaining components can reduce this burden partially, as can well-designed components for black-box reuse. But this will only be possible in well-understood domains. The generative approach imposes a different understanding cost, in learning the requirements language. If the language is already part of the users' or designers' technical vocabulary then these costs will be negligible, so end-user programming with generative architectures should be competitive in narrow technical communities. Implementation

Table 8.1 Relative costs for end-user programming and component engineering.

	Understanding	Implementation/ development	Initial investment	Payoff	Limitations
End-user programming	Requirements language	Expressing requirements	High: many generators	Technical and general domains	Narrow domains
Component engineering	Component design and functions	Design and test. Expressing requirements	High: component library	Technical domains	None if appropriate library exists

costs for end-user programming are minimal. Once the requirements have been specified, the system automatically creates the product. In contrast, component engineering always incurs some design and implementation costs, even in black-box reuse. However, these costs could be hidden in a hybrid approach that generates applications from a requirements language using a component library.

Initial investment costs are considerable in both approaches. Component engineering has to create the reuse library and support tools, whereas the generator approach has to develop the application generator architecture and environment. If the ideal of a general-purpose requirements language were achievable then economies of scale would favour this approach, but such an ideal is many years in the future. Practical application generators could be built for many separate and narrow domains, so there is an interesting economic balance between developing many generators for several domains, and component libraries with support tools that could cover a similar span of domains.

Component engineering may be seen by designers as abrogating their responsibility for design detail. This may clash with many designers' sense of professionalism. For end users this is not a concern. They just want to get a product that fits their needs as painlessly as possible. The generative approach appears to be the only acceptable way for most users, but it is likely to be difficult for several years to come. General-purpose requirements languages will be the semi-formal expressions of designs. In narrow technical user domains, sub-languages may be exploited more effectively, but this is no panacea for more open-ended domains. One prospect lies in the difference between software markets for individuals and for organizations. Corporate software procurement tends to be more conservative. Cost reduction is a constant driver but so is reliability. Reuse libraries, once established, offer better guarantees of reliability and considerable cost reduction over bespoke development. End users, in contrast, will be more sensitive to imposed costs and product price. Either they will choose software as they currently do in the COTS (customer off-the-shelf) market place, or more tractable requirements specification languages will evolve so application generation becomes a low cost (in cognitive and economic terms) alternative. However, I doubt this will happen in the near future. Natural language processing and knowledge acquisition for general-purpose reasoning are the barriers. In spite of these limitations there is still great potential for research into more generally applicable requirements languages, and application generator technology. The natural language community has spent much effort in inferring facts, arguments and even stories from natural language texts in restricted domains (Humphreys et al., 1999; Cowie and Wilks, 2000). This research needs to be integrated with software and requirements engineering. Initial attempts in this direction have solved the easy problems of deducing data structures and entity

relationship models from natural language texts (Black *et al.*, 1987). The more difficult problem of inferring functions, behaviour and goals from requirements
expressed in natural language has yet to be addressed.

8.7 Requirements in Systems Engineering

RE has maintained a somewhat schizophrenic relationship between requirements
analysis with software in mind (i.e. software systems requirements) and requirements that are applicable to large-scale socio-technical systems. Increasingly, systems are becoming mobile and ubiquitous; they link people together in increasingly
complex networks. System-level requirements raise the problem of how to specify
requirements for people as well as machines. In earlier chapters I reviewed some of
the research into functional allocation that has addressed this problem; however,
much more needs to be done. Interactive systems requirements involve deciding
how human tasks can be supported, and the design of human-computer collaboration. While HCI can provide task analysis methods that offer some guidance, there
is a pressing need to go beyond the simple functional allocation heuristics that have
existed for several years (see Sheridan, 2000). Some promising directions are defining requirements for generic task models in domains (e.g. command-and-control
IDAS model: Dearden *et al.*, 2000) and proposing generic tasks with associated
requirements issues and design rationale (Sutcliffe, 2002) that indicate functional
allocation decisions.

The old division between software and systems requirements is becoming
increasingly hard to defend. Yet the RE community has hardly engaged systems
engineers, and vice versa. Systems engineers in the INCOSE community view
requirements as complex models of socio-technical systems. Requirements are
understood as technical definitions of equipment performance that are measurable.
These may incorporate functional and non-functional requirements; for instance, a
radar system will be capable of detecting aircraft flying above 1000 metres at 100
miles range in a variety of weather conditions. The problem with system-level
requirements lies in specifying the range of operational conditions under which the
equipment is expected to operate. This has led to the concept of a necessary and sufficient set of scenarios that could be used to test a system model. Unfortunately, scenarios usually involve many environmental variables, most of which can vary
independently, so there is an exponential explosion of possible states the system
could encounter. Bounding the space of necessary and sufficient scenarios is a
pressing research problem in systems and requirements engineering.

8.8 Requirements and Software Evolution

RE has been aware of the moving target problem for many years, ever since Lehman
(1990) drew attention to classes of application in which requirements are so volatile
that they change before the software engineer has implemented the system. In business domains the pace of change accelerates inexorably. So far, RE has advocated
prototyping and rapid application development techniques (DSDM, 1995). In the
future we need to examine how requirements can be specified so software can
evolve and adapt to the customers' needs.

Software will always have a cost. It is knowledge- and people-intensive to construct. There are some approaches in the gift economy where software is freely distributed, Linux being a well-known example; however, it is unlikely that most software houses will follow Torvas Linvald's example. Like most products, software has a perceived utility that governs the price people are willing to pay for it. We can apply marketing theory to any software product and predict that a product's success will depend on:

- *price* – in relation to the perceived utility and competitor products;
- *population* – of purchasers of the product;
- *place* – the localization of the market, the spread of publicity and knowledge about the product;
- *promotion* – advertising and the level of education of potential customers about the merits of the product.

Clearly these factors interact; as knowledge of the product spreads via publicity, the market place becomes more educated, and more potential purchasers are available. Early entry products sometimes fail because of insufficient market education, leaving a successor product to reap the rewards (e.g. the Xerox Star workstation which Apple capitalized on with the Macintosh). Early products tend to compete on functionality and features and have few competitors, whereas more mature products face many competitors, compete more on usability than functionality, and have lower price margins (Norman, 1999). This digression into marketing theory has a point. The development approach may depend on the maturity of the product. Innovative applications will always be designed *de novo*. They are unlikely to be developed by the generative approach because it would take longer to develop the generator for requirements in a new domain than it would to develop the product from scratch. Another implication is the impact of perceived utility and the importance of usability. Users in immature markets tend to have higher motivation to try out exciting new functionality, particularly early adopters, so they will put up with design and customization burdens. The fit between the users' need and product functionality has more tolerance. This is less true in mature markets where supplier reputations have developed and users place more value on usability and an exact fit of functions to their needs.

Therefore we need to consider not only costs of the product but also costs of adoption by the user. Technology acceptance models (Davis, 1993), based on the theory of adaptive decision-making (Ajzen and Fishbein, 1980), assert that purchasing (or any other) decisions are governed by weighing the merits of attributes of rival products. Requirements for COTS selection or product procurement will be matched against these attributes, so we need to review just what those attributes may be:

- *Perceived utility* – the expected value to the individual in saving time, increasing performance or quality, and achieving the user's goal for work-oriented software; for entertainment applications this will be the perceived potential for excitement and fun.
- *Actual utility* – the value experienced by the individual accumulated over time, which may be very different from expectations.
- *Usability* – ease of learning and ease of use, which will be more important in mature markets.

- *Cost of purchase* – the acceptable cost will be influenced by the perceived and actual utility and the individual's (or organization's) relative wealth.
- *Brand identity and reputation* – purchasers often prefer a reputable source.
- *Convenience* – ease of set up and time to effect use.

These variables will have complex relationships. For example, Media Communication Theory (Daft, Lengel and Trevino, 1987) points out that information about products can be obtained from broadcast or narrowcast sources. Advertising is broadcast whereas word-of-mouth recommendations deliver targeted advice. Product surveys and reference sites fit in between. Brand identity, trust and perceived utility may be acquired from several sources and the importance attached to this information will vary according to the users' trust of the source. The transformation of perceived utility into actual utility depends on experience. But experience may have a history. If the customer has experience of similar products from the same supplier, and those experiences have been positive, then trust will have been created. Moreover this will have a positive effect on perceived utility for the next generation of products. Conversely if the experience was negative then trust and perceived utility will be affected adversely. So the success of an application will depend on a complex trade-off between the classic view of functional requirements being satisfied by a design, but modified by trust in the supplier and the expected effort in acquiring or building the desired product. Expected effort may reflect the trade-off between understanding a requirements language or a design support tool or COTS selection set against the desired degree of satisfaction of requirements achieved by a design.

Purchasing decisions will be a function of cost-benefit trade-offs. Cost will be incurred by a less than perfect task fit for the customer population. Design and configuration effort are incurred to improve the task fit, cost of failures (software reliability and usability errors), and convenience costs in set up. Set against these costs, the benefit side of the equation is composed of perceived utility, plus actual utility for repeat purchases, and values established by trust of the supplier and brand identity. These can be summarized as follows:

Value f . (PercUt +Brand) – Costs (Design + Task-fit + Usab + Reliab + Conv + Errors)

Acceptability and purchase will depend on the value/purchase price balance for any one user or organization. So if brand identity and perceived utility are low, lowering design, usability and convenience costs will not dramatically increase the relative value, and purchase is less likely. Not only do requirements and how they are achieved in a design need to be advertised in the traditional sense, but also software in the future may need to advertise itself, especially if it is designed for reuse. In the future, requirements for explanation, promotion and attractiveness will become more prominent as software needs more than a sound implementation of functional requirements to succeed in the market place. Clearly the value placed on utility, usability, attractiveness and so on, will be different in corporate and individual purchasing decisions. Individuals will value usability and convenience more whereas managers may (but shouldn't) place more weight on utility. In mature markets, users will place more weight on the costs so products will have to minimize these by better usability, reliability and convenience with excellent task fit and low design costs.

Another factor in the purchasing equation is reputation. This is the collected wisdom of other users that relates to usability and actual utility. As the Internet lowers the costs of spreading information in electronic fora and catalogues, reputation information will feed back more rapidly and poor products will become eliminated. Adaptable products, in theory, should outscore other approaches, as they could ideally adapt their functionality to suit the users' needs, improve usability by adapting to users' characteristics, and reduce convenience costs by auto-configuration. Unfortunately, adaptable products often make mistakes in interpreting users' requirements, whether input as a sub-language or inferred from user behaviour. The sting of high error costs makes adaptation a risky strategy. Finally, purchase and product success will vary over time, as illustrated in Figure 8.3.

In products created by reuse there is a high initial cost of creating the reuse library. This has to be set against the expected return later in the life cycle when design by reuse is realized. In immature markets, however, products are adopted by only a few pioneer users, so returns are initially low. If the product survives then more people adopt it and the return on investment grows; finally the market becomes mature and the majority adopt the products, but by then many competitors are usually present, so profits may not be so high. Designing innovative reuse libraries is difficult because two burdens have to be overcome, the initial investment and the low return from an immature market; consequently reuse is more likely to spread in maturing markets where the return on investment is more likely to be realized.

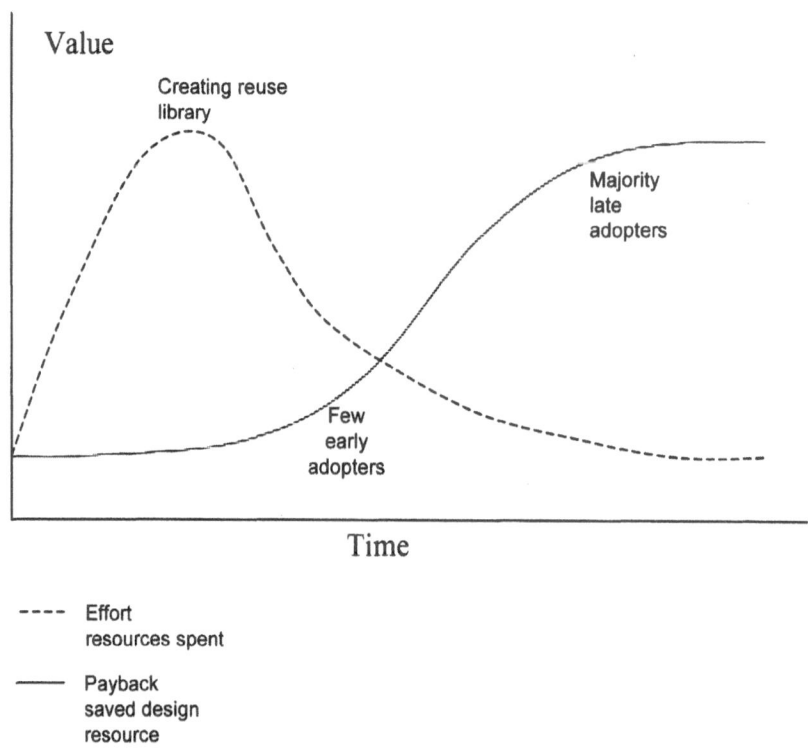

Fig. 8.3 Effort and payback in reuse-led development.

8.9 RE Challenges

Michael Jackson's conception of requirements engineering being the dependency between properties of a design machine, the designer's intentions within the context of a domain and its properties can be considered as a nascent theory of RE. So what are the prospects for development of such a theory? While Jackson's argument is a useful starting point it encounters problems of complexity and scale in the real world. The problems are twofold. First is the knowability problem of domain properties that may influence the dependency between the requirements and the design specification that should satisfy them. In systems with limited interaction with the external world or people, the dependencies and the domain can be understood and modelled in detail. For instance, the specification of a car cruise control system can be formally specified so it will work accurately within the limits of the known domain. The problem arises with biddable domains, as Jackson calls them. People are biddable domains, and their behaviour can only be modelled in probabilistic terms, and even then with limited accuracy (Galliers *et al.*, 1999).

To illustrate the point with the cruise control, assuming we had specified a correct design that satisfied the basic requirements of setting the speed, controlling the car, and interrupts for the driver to regain control, have we anticipated all the possible effects of such a designed system? If the cruise control actually brakes the car going downhill to maintain constant speed, this braking is dangerous because drivers tailgate on motorways and normally we expect cars to accelerate going downhill, so collisions may result. We have discovered emergent requirements that came from the social domain of experience. While the cruise control example may not be too serious, when we are specifying requirements for complex safety critical software such as fly-by-wire avionics, the implications are far more serious. Several airbus A320 accidents have been caused by pilots behaving in a manner that the design did not anticipate. RE will have to deal with knowability, or unpredictability, of human behaviour. Only when a system is implemented do the consequences of its use become apparent. While 20/20 foresight is always going to be an impossible goal, RE does have a duty to anticipate the effects of design as far as possible. We have developed methods and tools that help to capture requirements, and can reason about the behaviour of design machines in the software world. Unfortunately that leaves the external world and many domain properties in a largely unknown state. To make progress, RE will have to develop probabilistic models that can predict behaviour in the messy, uncertain, human world so that requirements can be validated with more certainty. This will be no easy endeavour. The number of variables and their combination presents a huge problem space. People even in a general sense are governed by their ability, knowledge and motivation, as well as being affected by fatigue and stress. Human behaviour is influenced by many environmental variables such as noise, heat, humidity, climate and weather.

This discussion has only started to explore the problem, so one may doubt that this direction is tractable. Only further research will tell, but there are some modest beginnings. The predictability of human behaviour when interacting with requirements/designs has been a concern of researchers in safety critical systems (Reason, 1997; Hollnagel and Bye, 2000) who have created taxonomies of influencing factors. Simple models have been produced to predict human errors that can account for designers, albeit at a crude level (Galliers *et al.*, 1999; Hollnagel and Bye, 2000). In the future we will need families of predictive models to help validate requirements from different viewpoints, for example human errors and mistakes, problems in social organization, trust and persuasion in technology, etc.

So where is the future for RE? One reflection on the future is how machine learning may play a role. The difficulty of customizing and adapting designs to users' requirements has already been noted, but that does not mean the problem is insoluble. Building general-purpose learning machines is just as difficult as natural language understanding. These problems are just different sides of the same artificial intelligence coin. To understand natural language you need to acquire large quantities of complex knowledge. To acquire complex knowledge takes time, and the only efficient way is to build learning processes. Acquiring requirements is a form of knowledge acquisition, so if applications could learn and adapt to new circumstances then this could solve the problem.

Some first steps have been taken in this direction in the concept of requirements monitoring (Fickas and Feather, 1995). If an application can detect changes in its environment and has an embedded model that enables it to interpret those changes, and furthermore the ability to infer the implications of those changes for design, then it could auto-adapt its behaviour for evolving requirements. This vision is not new. The human–computer interaction community proposed adaptive and adaptable systems some time ago (Browne, Totterdell and Norman, 1990), but progress has been slow. Unfortunately there is no simple silver bullet. Learning machines require embedded models of the world to improve efficiency of learning, as well as more sophisticated learning algorithms to infer deep knowledge for more surface-level information. While general-purpose learning machines are still an open research issue, more humble adaptable components with limited ability to learn are feasible.

A means to this end is to apply the principle of Darwinian evolution to the problem. If software is to evolve then it should live or die by the laws of natural selection. If applications can monitor their environment, implement and detect changes in the environment, then with suitable rules it should be possible to make the designed application adapt automatically to its changed circumstances. This becomes requirements engineering to specify adaptive software, but it still necessitates that the initial design is specified in the first place. However, many design problems involve optimizations, trade-offs and matching constraints. Evolutionary computing could provide the answer by generating many different design permutations automatically and then letting a selection mechanism find the best option to fit the trade-off criteria. In this approach the requirements engineer just sets the bounds of the problem, specifies the criteria for a successful design (probably NFRs), seeds the environment with some design components and then lets the design environment do the rest.

These concepts have been applied to software in the form of genetic programming (Back, 1997). A population of slightly different programs is created and then run to create outputs. Survivorship criteria are applied to select only those variants that had optimal or acceptable results. The population is then allowed to evolve by adopting more lessons from biology. Programs can mutate in a semi-random fashion to create new variables, actions or possible conditions. Crossover allows programs to exchange different segments, such as conditional tests or operations. The new population is rerun against the survival criteria and variants that compete successfully survive to the next generation. Evolutionary computing has been successfully applied for deriving optimal solutions for a narrow range of algorithm optimizations. There is an opportunity to apply such ideas to the world of requirements.

The final vision for an evolving software environment is shown in Figure 8.4.

First-generation designs would be run in an environment which imposes requirements as environmental constraints and sets performance criteria. The initial population of designs is reduced by (artificial) natural selection. The next generation of designs is created automatically by crossover of components between different slots in the application template, or by semi-random mutation to generate new components from the library. The mutation, crossover process would have to be controlled by rules that constrain allocation of component types to particular template slots, but these rules are already part of many application framework architectures (Fayad and Johnson, 2000). Over many runs with careful design of crossover and selection rules the design population should stabilize with an optimal solution. The next step for software evolution is to make the domain environment capable of detecting and interpreting changes. It then changes its own survivorship parameters and triggers further design evolutions to fit with the changed requirements. This approach will not solve all reuse problems. For instance, composition rules and the evolutionary mechanism will still have to be specified. Furthermore, the environment will have to be limited in scope so that a tractable set of survivorship parameters and composition rules can be specified. This approach is currently being researched in the ERSRC SIMP project for requirements evolution, with a restricted set of command and control applications.

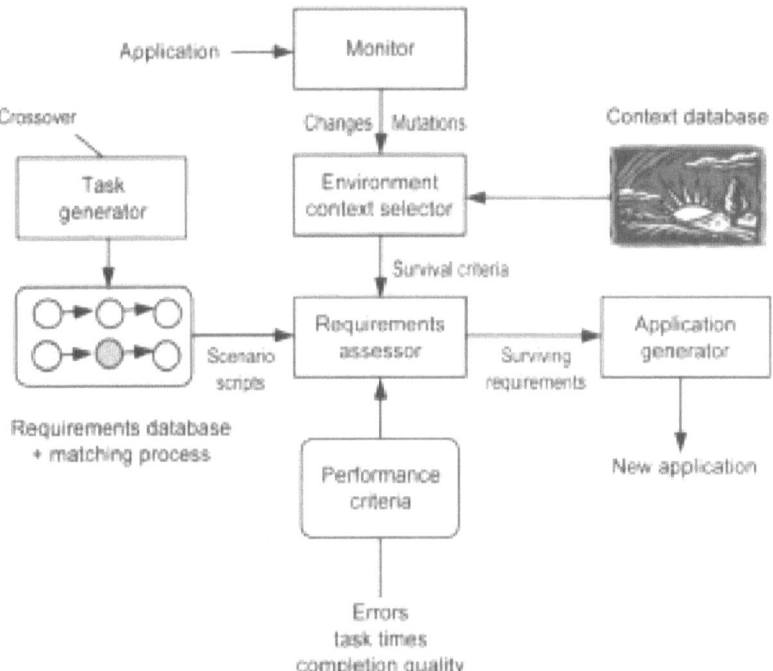

Fig. 8.4 Software evolution environment.

8.10 Summary

The prospects for the future of RE are investigated. Three separate pathways may develop in the future. First is requirements for reuse-driven design and component engineering. This is already gaining ground in ERPs and application frameworks. Second is application generators coupled to high-level requirements languages. Even though high-level requirements languages have not had much commercial impact so far, domain specific sub-languages and even general purpose near-natural English requirements languages may play an increasingly important role in the future. The third avenue is to pass the responsibility for requirements capture and design on to the end user. End-user development environments equipped with high-level tools could allow users to build applications by a mixture of demonstration and more explicit design. However, many users may not wish to accept the outsourcing of design, so requirements engineers will be needed for many years to come. Furthermore, many aspects of RE involve communication, judgement and human sensitivity for handling negotiations and trade-offs between different stakeholder requirements.

Adaptation appears to be a risky strategy because of the potential cost of errors and maladaptation. Component engineering and application generation have lower costs but application generators will be more suitable for end users, while component engineering may prove to be acceptable for technical users and large organizations. A model of future intelligent reusable components is proposed with an application generator architecture that assembles applications from semi-autonomous agents. However, this poses problems of distributing knowledge among agents, so a final vision for software evolution is to apply genetic algorithm approaches. An evolutionary application environment senses the world and sets survivorship criteria for evolution populations of applications. Both compositional approaches to reuse and requirements languages with generative architectures encounter the Tower of Babel problem. Maybe standardization will take hold, or perhaps evolutionary programming and intelligent generative architectures may produce the eventual solution.

References

Ajzen, I and Fishbein, M (1980) *Understanding attitudes and predicting social behaviour*. Prentice Hall, Englewood Cliffs NJ.

Anderson, JR (1985) *Cognitive psychology and its implications*. Freeman, New York.

Anderson, S and Tourlas, K (1998) Design for proof: An approach to the design of domain specific languages. *Formal Aspects of Computing* 10, pp. 452–68.

Anton, AI and Potts, C (1998) The use of goals to surface requirements for evolving systems. In: *Proceedings 1998 International Conference on Software Engineering: Forging New Links*, Kyoto, Japan. IEEE Computer Society Press, Los Alamitos CA, pp. 157–66.

Arens, Y, Hovy, E and Vossers, M (1993) On the knowledge underlying multimedia presentations. In: Maybury, MT (ed.) *Intelligent multimedia interfaces*. AAAI/MIT Press, Cambridge MA, pp. 280–306.

Attwood, ME, Burns, B, Girgensohn, A, Lee, A, Turner, T and Zimmerman, B (1995) Prototyping considered dangerous. In: Nordby, K, Helmersen, PH, Gilmore, DJ and Arnesen, SA (eds) *Proceedings Human Computer Interaction INTERACT 95*. IFIP/Chapman and Hall, London, pp. 179–84.

Austin, JL (1962) *How to do things with words*. Oxford University Press, London.

Back, T (ed.) (1997) *Handbook of evolutionary computation*. Institute of Physics Publishing/Oxford University Press, Oxford.

Bailey, RW (1982) *Human performance engineering: A guide for system designers*. Prentice Hall, Englewood Cliffs NJ.

Batory, D (1998) Domain analysis for GenVoca generators. In: *Proceedings 5th International Conference on Software Reuse*. IEEE Computer Society Press, Los Alamitos CA, pp. 350–51.

Beer, S (1981) *The brain of the firm*. Wiley, Chichester.

Bell, TE and Thayer, TA (1976) Software requirements: Are they really a problem? In: *Proceedings 2nd International Conference on Software Engineering*. IEEE Computer Society Press, Los Alamitos CA, pp. 61–68.

Beyer, H and Holtzblatt, K (1998) *Contextual design: Defining customer-centered systems*. Morgan Kaufmann, San Francisco.

Black, WJ, Sutcliffe, AG, Loucopoulos, P and Layzell, PJ (1987) Translation between pragmatic software development methods. In: Nichols, HK and Simpson, D (eds) *Proceedings of the 1st European Conference on Software Engineering*, Strasbourg. Springer-Verlag, Berlin.

Boehm, B (1981) *Software engineering economics*. Prentice Hall, Englewood Cliffs NJ.

Boehm, B, Bose, P, Horowitz, E and Lee, MJ (1994) Software requirements as negotiated win conditions. In: *Proceedings Requirements Engineering 94*. IEEE Computer Society Press, Los Alamitos CA, pp. 74–83.

Bowers, J, Viller, S and Rhodden, T (1994) Human factors in requirements engineering, Department of Computer Science Report No. REAIMS Deliverable D1. University of Lancaster, Lancaster.

Bradach, JL and Eccles, RG (1989) Price authority and trust: From ideal types to plural forms. *Annual Review of Sociology* 15, pp. 97–118.

Brereton, P, Budgen, D, Bennett, K, Munro, M, Layzell, P, Macaulay, L, Griffiths, D and Stannett, C (1999) The future of software: Defining the research agenda. *Communications of the ACM* 42, pp. 78–84.

Breuker, J and Van Der Velde, W (1994) *CommonKADS Library for expertise modelling*. IOS Press, Amsterdam.

Briand, L, Melo, W, Seaman, C and Basili, V (1995) Characterizing and assessing a large-scale software maintenance organization. In: Jeffrey, R and Notkin, D (eds) *Proceedings 17th International Conference on Software Engineering (ICSE 17)*, Seattle. IEEE Computer Society Press, Los Alamitos CA, pp. 133–43.

Brown, P and Levinson, SC (1987) *Politeness: Some universals in language usage.* Cambridge University Press, Cambridge.

Browne, DP, Totterdell, P and Norman MA (eds) (1990) *Adaptive user interfaces.* Academic Press, London.

Bubenko, J (1993) Extending the scope of information modelling. In: *Proceedings 4th International Workshop on the Deductive Approach to Information Systems and Databases* [available as SISU Research Report DSV 93-034], Lloret. SISU, Stockholm.

Buckingham Shum, S (1996) Analyzing the usability of a design rationale notation. In: Moran, TP and Carroll, JM (eds) *Design rationale: Concepts, techniques and use.* Lawrence Erlbaum Associates, Hillsdale NJ, pp. 185–215.

Cacioppo, JT, Petty, RE and Kao, CF (1984) The efficient assessment of need for cognition. *Journal of Personality Assessment* 48, pp. 306–307.

Card, SK, Moran, TP and Newell, A (1983) *The psychology of human computer interaction.* Lawrence Erlbaum Associates, Hillsdale NJ.

Carroll, JM (2000) *Making use: Scenario-based design of human-computer interactions.* MIT Press, Cambridge MA.

Carroll, JM (ed.) (1995) *Scenario-based design: Envisioning work and technology in system development.* Wiley, New York.

Carroll, JM and Rosson, MB (1992) Getting around the task-artifact framework: How to make claims and design by scenario. *ACM Transactions on Information Systems* 10, pp. 181–212.

Carroll, JM, Alpert, SR, Karat, J, Van Deusen, MD and Rosson, MB (1994a) Raison d'être: Capturing design history and rationale in multimedia narratives. In: Adelson, B, Dumais, S and Olson, J (eds) *Human Factors in Computing Systems: CHI 94 Conference Proceedings.* ACM Press, New York, pp. 192–97.

Carroll, JM, Alpert, SR, Karat, J, Van Deusen, MD and Rosson, MB (1994b) Raison d'être: Embodying design history and rationale in hypermedia folklore, an experiment in reflective design practice. *Library Hi-Tech* 12, pp. 59–70 and 81.

Checkland, P (1981) *Systems thinking, systems practice.* Wiley, Chichester.

Checkland, P and Scholes, J (1990) *Soft systems methodology in action.* Wiley, Chichester.

Chemical Industries Association (1993) A guide to hazard and operability studies. Chemical Industries Association, London.

Chung, L (1993) Representing and using non-functional requirements: A process-oriented approach. Research in data and knowledge base systems No. DKBS-TR-93-1. Department of Computer Science. University of Toronto, Toronto.

Clark, HH (1996) *Using language.* Cambridge University Press, Cambridge.

Coad, P, Yourdon, EE (1991) *Object-oriented analysis.* Yourdon Press, Englewood Cliffs, NJ.

Cockburn, A (2001) *Writing effective use cases.* Addison-Wesley, Boston, MA.

Conklin, J and Begeman, ML (1988) gIBIS: A hypertext tool for exploratory policy discussion. *ACM Transactions on Office Information Systems* 64, pp. 303–31.

Cowie, J, Wilks, Y (2000) Information extraction. In: Dale, R, Moisl, H and Summers, H (eds) *Handbook of natural language processing.* Marcel Dekker, New York.

Crinnion, J (1991) *Evolutionary systems development: A practical guide to the use of prototyping within a structured systems methodology.* Pitman, London.

Daft, RL, Lengel, RH and Trevino, LK (1987) Message equivocality, media selection, and manager performance: Implications for information systems. *MIS Quarterly* 11, pp. 355–66.

Dardenne, A, Van Lamsweerde, A and Fickas, S (1993) Goal directed requirements acquisition. *Science of Computer Programming* 20, pp. 3–50.

Davenport, T (1993) *Process innovation: Re-engineering work through information technology.* Harvard Business School Press, Boston, MA.

Davies, AM and Hsai, P (eds) (1994) *Proceedings IEEE International Conference on Requirements Engineering, Colorado Springs CO.* IEEE Computer Society Press, Los Alamitos CA.

Davis, FD (1993) User acceptance of information technology: System characteristics, user perceptions and behavioral impacts. *International Journal of Man–Machine Studies* 38, pp. 475–87.

De Marco, T (1978) *Structured analysis and systems specification.* Prentice Hall, Englewood Cliffs, NJ.

Dearden, A, Harrison, M and Wright, P (2000) Allocation of function: Scenarios, context and the economies of effort. *International Journal of Human–Computer Studies* 52, pp. 289–318.

Diller, A (1994) *Z: An introduction to formal methods.* Wiley, Chichester.

Doney, PM, Cannon, JP and Mullen, MR (1998) Understanding the influence of national culture on the development of trust. *Academy of Management Review* 23, pp. 601–21.

Downs, E, Clare, P and Coe, I (1992) *Structured systems analysis and design method: Application and context.* 2nd edn. Prentice Hall, New York.

DSDM (1995) *DSDM Consortium: Dynamic Systems Development Method.* Tesseract Publishers, Farnham Surrey.

Dubois, E, Hagelstein, J and Rifaut, A (1989) Formal requirements engineering with ERAE. *Philips Journal of Research* 43.

Dubois, P, Dubois, E and Zeippen, J (1997) On the use of a formal representation. In: *Proceedings ISRE 97: 3rd IEEE International Symposium on Requirements Engineering,* Annapolis MD. IEEE Computer Society Press, Los Alamitos CA, pp. 128–37.

Easterbrook, S and Callahan, J (1997) Formal methods for V&V of partial specifications: An experience report. In: Heitmeyer, C and Mylopoulos, J (eds) *Proceedings Third IEEE International Symposium of Requirements Engineering.* IEEE Computer Society Press, Los Alamitos CA, pp. 160–68.

Eason, KD (1988) *Information technology and social change.* Taylor and Francis, London.

Eden, C (1988) Cognitive mapping. *European Journal of Operational Research,* pp. 1–13.

Edwards, MD (1995) Hardware/software co-design: Experiences with languages and architectures. In: Buchenrieder, K and Rosenblit, J (eds) *CODESIGN computer-aided software/hardware engineering.* IEEE Computer Society Press, Los Alamitos CA, pp. 356–77.

El Emam, K and Madhavji, NH (1995) A field study of requirements engineering practices in information systems development. In: Harrison, MD, Zave, P (eds) *Proceedings 1995 IEEE International Symposium on Requirements Engineering (RE 95),* York. IEEE Computer Society Press, Los Alamitos CA, pp. 68–80.

Ellyson, SL and Dovidio, JF (1985) *Power dominance and non-verbal behaviour.* Springer-Verlag, New York.

Emerson, R (1962) Power dependence relations. *American Sociological Review* 27, pp. 31–41.

Fayad, ME and Johnson, RE (2000) *Domain-specific application frameworks: Frameworks experience by industry.* Wiley, New York.

Fenton, NE (1995) The role of measurement in software safety assessment. In: *Proceedings CSR/ENCRESS Conference,* Bruges.

Fickas, S and Feather, MS (1995) Requirements monitoring in dynamic environments. In: Harrison, MD and Zave, P (eds) *Proceedings 1995 IEEE International Symposium on Requirements Engineering (RE 95),* York. IEEE Computer Society Press, Los Alamitos CA, pp. 140–47.

Finkelstein, ACW and Dowell, J (1996) A comedy of errors: The London Ambulance Service case study. In: *Proceedings 8th International Workshop on Software Specification and Design,* Schloss Velen, Germany. IEEE Computer Society Press, Los Alamitos CA, pp. 2–4.

Finkelstein, ACW and Fickas, S (eds) (1993) *Proceedings of IEEE Symposium on Requirements Engineering.* IEEE Computer Society Press, Los Alamitos CA.

Finkelstein, ACW, Kramer, J and Nuseibeh, B (1992) Viewpoints: A framework for integrating multiple perspectives in system development. *International Journal of Software Engineering and Knowledge Engineering* 2, pp. 31–57.

Fischer, G (1996) Seeding evolutionary growth and reseeding: Constructing, capturing and evolving knowledge in domain-oriented design environments. In: Sutcliffe, AG, Benyon, D and Van Assche, F (eds) *Domain knowledge for interactive system design: Proceedings TC8/WG82 Conference on Domain Knowledge in Interactive System Design.* Chapman and Hall, London, pp. 1–16.

Fischer, G, Girensohn, A, Nakakoji, K and Redmiles, D (1992) Supporting software designers with integrated domain-oriented design environments. *IEEE Transactions on Software Engineering* 18, pp. 511–22.

Fischer, G, Nakakoji, K, Otswald, J, Stahl, G and Summer, T (1993) Embedding computer-based critics in the contexts of design. In: Ashlund, S, Mullet, K, Henderson, A, Hollnagel E and White, T (eds) *Human Factors in Computing Systems: INTERCHI 93 Conference Proceedings.* ACM Press, New York, pp. 157–63.

Fisher, S (1994) *Multimedia authoring: Building and developing documents.* AP Professional, Cambridge, MA.

Fogg, BJ (1998) Persuasive computer: Perspectives and research directions. In: *Human Factors in Computing Systems: CHI 98 Conference Proceedings.* ACM Press, New York, pp. 225–32.

Freeman, PA (1987) A conceptual analysis of the Draco approach to constructing software systems. *IEEE Transactions on Software Engineering* 13, pp. 830–44.

Fukuyama, F (1995) *Trust: The social virtues and the creation of prosperity.* Free Press, New York.

Galliers, J, Sutcliffe, AG and Minocha, S (1999) An impact analysis method for safety-critical user interface design. *ACM Transactions on Computer–Human Interaction* 6, pp. 341–69.

Gans, G, Jarke, M, Kethers, S, Lakemeyer, G, Ellrich, L, Funken, G and Meister, M (2001) Requirements modelling for organisation networks: A (dis) trust-based approach. In: Easterbrook, S and Nuseibeh, B (eds) *Proceedings RE 01: 5th IEEE International Symposium on Requirements Engineering*. IEEE Computer Society Press, Los Alamitos CA, pp. 154–63.

Gardiner, M and Christie, B (eds) (1987) *Applying cognitive psychology to user interface design*. Wiley, Chichester.

Gaskell, C and Phillips, R (1995) A structured analysis formalism with execution semantics to allow unambiguous model interpretation. In: Schaeffer, W and Botella, P (eds) *Proceedings of Software Engineering ESEC 95, 5th European Software Engineering Conference* (Springer-Verlag Lecture Notes in Computer Science, 989). Springer-Verlag, Berlin, pp. 235–53.

Gause, DC and Weinberg, GM (1989) *Exploring requirements: Quality before design*. Dorset House, New York.

Gentner, D and Stevens, AL (eds) (1983) *Mental models*. Lawrence Erlbaum Associates, Hillsdale, NJ.

Gibson, JJ (1986) *The ecological approach to visual perception*. Lawrence Erlbaum Associates, Hillsdale, NJ.

Goguen, J (1993) Social issues in requirements engineering. In: *Proceedings 1st International Symposium on Requirements Engineering*. IEEE Computer Society Press, Los Alamitos CA, pp. 194–95.

Goguen, J and Linde, C (1993) Techniques for requirements elicitation. In: Fickas, S and Finkelstein, ACW (eds) *Proceedings 1st International Symposium on Requirements Engineering*. IEEE Computer Society Press, Los Alamitos CA, pp. 152–64.

Goldin, L, Berry, DM, Sutcliffe, AG and Ryan, K (1997) AbstFinder, a prototype natural language text abstraction finder for use in requirements elicitation. *Automated Software Engineering* 4, pp. 375–418.

Gotel, OTZ and Finkelstein, ACW (1994) An analysis of the requirements traceability problem. In: *Proceedings IEEE International Conference on Requirements Engineering*. IEEE Computer Society Press, Los Alamitos CA, pp. 94–101.

Gould, JD (1987) How to design usable systems. In: Bullinger, HJ and Shackel, B (eds) *Proceedings Second IFIP Conference on Human–Computer Interaction, Interact87*, Stuttgart. North-Holland, Amsterdam, pp. xxxv–xxxix.

Graham, L (1998) *Principles of interactive design*. Delmar, Albany NY.

Green, TRG and Petre, M (1996) Usability analysis of visual programming environments: A cognitive dimensions framework. *Journal of Visual Languages and Computing* 7, pp. 131–74.

Greenspan, S, Borgida A and Mylopoulos, J (1986) A requirements modelling language and its logic. *Information Systems* 11, pp. 9–23.

Grice, HP (1975) Logic and conversation. *Syntax and Semantics* 3.

Grishman, R and Kittredge, R (eds) (1986) *Analysing language in restricted domains: Sub-language description and processing*. Lawrence Erlbaum Associates, Hillsdale NJ.

Griss, ML, Favaro, J and d'Alessandro, M (1998) Integrating feature modeling with the RSEB. In: Devanbu, P and Poulin, J (eds) *Proceedings Fifth International Conference on Software Reuse*, Victoria BC. IEEE Computer Society Press, Los Alamitos CA, pp. 76–85.

Grudin, J (1991) Systematic sources of sub-optimal interface design in large product development organisations. *Human–Computer Interaction* 6, pp. 147–96.

Hammer, M and Champy, J (1993) *Re-engineering the corporation: A manifesto for business revolution*. Nicholas Brealy, London.

Harel, D (1987) Statecharts: a visual formalism for complex systems. *Scientific Computer Programming* 8, pp. 231–74.

Harker, SDP, Eason, KD and Dobson, JE (1993) The change and evolution of requirements as a challenge to the practice of software engineering. In: Fickas, S and Finkelstein, ACW (eds) *Proceedings 1st International Symposium on Requirements Engineering*. IEEE Computer Society Press, Los Alamitos CA, pp. 266–72.

Harris, ZS (1989) *The form of information in science: Analysis of an immunology sub-language*. Kluwer Academic Publishers, Dordrecht.

Hart, SG and Staveland, LE (1988) Development of a NASA-TLX (Task Load Index): Results of empirical and theoretical research. In: Hancock, PS and Meshkati, N (eds) *Human mental workload*. Elsevier, Amsterdam, pp. 139–83.

Haumer, P, Pohl, K and Weidenhaupt, K (1998) Requirements elicitation and validation with real world scenes. *IEEE Transactions on Software Engineering* 24, pp. 1036–54.

Hauser, J and Clausing, D (1988) The house of quality. *Harvard Business Review* 5, pp. 63–73.

Heath, C and Luff, P (1992) Collaboration and control: Crisis management and multimedia technology in London Underground control rooms. *Computer Supported Co-operative Work* 1, pp. 69–94.

Heitmeyer, CL, Jeffords, RD and Labaw, BG (1996) Automated consistency checking of requirements specifications. *ACM Transactions on Software Engineering and Methodology* 5, pp. 231–61.

Hendry, I (1995) *Human resource management: A strategic approach to employment*. Butterworth-Heinemann, Oxford.

Hix, D and Hartson, HR (1993) *Developing user interfaces: Ensuring usability through product and process*. Wiley, New York.

HMSO (1993) *Report of the Inquiry into the London Ambulance Service*. HMSO, London.

Holland, CP (1995) Co-operative supply chain management: The impact of interorganisation information systems. *Journal of Strategic Information Systems* 4, pp. 117–33.

Hollnagel, E (1993) *Human reliability analysis: Context and control*. Academic Press, London.

Hollnagel, E and Bye, A (2000) Principles for modelling function allocation. *International Journal of Human–Computer Studies* 52, pp. 253–65.

Hughes, J, O'Brien, J, Rhodden, T, Rouncefield, M and Sommerville, I (1995) Presenting ethnography in the requirements process. In: Zave, P, and Harrison, MD (eds) *Proceedings of RE 95 Second International Symposium on Requirements Engineering*, IEEE Computer Society Press, Los Alamitos CA, pp. 27–34.

Humphreys, K, Gaizauskas, R, Azzan, S, Huyck, C, Mitchell, B, Cunningham, H and Wilks, Y (1999) Description of the University of Sheffield LaSIE II system as used for MUC-7. In: *Proceedings 7th Message Understanding Conference*. Morgan Kaufmann, San Francisco.

IEEE-TSE (1991) Special issue on requirements engineering. *IEEE Transactions on Software Engineering* 17.

IEEE-TSE (1992) Special issue on requirements engineering. *IEEE Transactions on Software Engineering* 18.

ISO (1998) ISO 14915 Multimedia user interface design software ergonomic requirements, Part 1: Introduction and framework; Part 3: Media combination and selection. International Standards Organisation.

Jackson, M (1983) *Systems development*. Prentice Hall, Hemel Hempstead.

Jackson, M (1995) *Software requirements and specifications: A lexicon of practice, principles and prejudices*. Addison-Wesley, Wokingham.

Jackson, M and Zave, P (1993) Domain descriptions. In: Fickas, S and Finkelstein ACW (eds) *Proceedings 1st International Symposium on Requirements Engineering – RE93*, San Diego CA. IEEE Computer Society Press, Los Alamitos CA, pp. 56–64.

Jacobs, S and Kethers, S (1994) Improving communication and decision making within quality function deployment. In: *Proceedings 1st International Conference on Concurrent Engineering, Research and Application*, Pittsburgh.

Jacobson, I (1987) Object oriented development in an industrial environment. In: *Proceedings OOPSLA 87 Conference on Object Oriented Programming, Systems, Languages and Applications, Orlando FL*. ACM Press, New York, pp. 183–91.

Jacobson, I, Christerson, M, Jonsson, P and Overgaard, G (1992) *Object-oriented software engineering: A use-case driven approach*. Addison Wesley, Reading, MA.

Jarke, M, Pohl, K, Sutcliffe, AG *et al.* (1993) Requirements engineering: An integrated view of representation. In: Sommerville, I and Manfred, P (eds) *Proceedings 4th European Software Engineering Conference*, Garmisch-Partenkirchen. Springer-Verlag, Berlin, pp. 100–14.

Johnson, CW (1996) Documenting the design of safety critical user interfaces. *Human–Computer Interaction* 8, pp. 221–39.

Johnson, CW (1999) Taking fun seriously: Using cognitive models to reason about interaction with computer games. *Personal technologies* 3, pp. 105–16.

Johnson, P (1992) *Human computer interaction: Psychology, task analysis and software engineering*. McGraw-Hill, London.

Johnson, WL, Feather, MS and Harris, DR (1992) Representation and presentation of requirements knowledge. *IEEE Transactions on Software Engineering* 18, pp. 853–69.

Johnson-Laird, PN (1983) *Mental models: Towards a cognitive science of language, inference and consciousness*. Cambridge University Press, Cambridge

Johnson-Laird, PN (1988) *The computer and the mind: An introduction to cognitive science*. Harvard University Press, Cambridge, MA.

Johnson-Laird, PN and Wason, PC (1983) *Thinking: Readings in cognitive science*. Cambridge University Press, Cambridge,

Kahnemann, D and Tversky, A (1982) Intuitive prediction: Biases and corrective procedures. In: Kahnemann, D, Slovic, P and Tversky, A (eds) *Judgement under uncertainty: Heuristics and biases.* Cambridge University Press, Cambridge.

Keller, SE, Kahn, LG and Parna, RB (1990) Specifying software quality requirements with metrics: Tutorial paper. In: Thayer, RH and Dorfman, M (eds) *System and software requirements engineering.* IEEE Computer Society Press, Los Alamitos CA, pp. 145–63.

Kelly, GA (1963) *A theory of personality.* WW Norton, New York.

Kennedy-Carter (2000) *Action specification language: Executable UML.* http://www.kc.com (accessed November 2000).

Kieras, DE and Polson, PG (1985) An approach to the formal analysis of user complexity. *International Journal of Man–Machine Studies* 22, pp. 365–94.

Kirikova, M and Bubenko, JA (1994) Enterprise modelling: Improving the quality of requirements specifications. Information systems research seminar No. IRIS-17, Olou, Finland.

Kobayaski, S (1971) *Creative management.* American Management Association, New York.

Kotonya, G and Sommerville, I (1996) Requirements engineering with viewpoints. *Software Engineering Journal* 11, pp. 5–18.

Krumbholz, M (1999) The impact of largescale ERP implementations on organisaions with diverse corporate, business and national cultures. In: *Proceedings Doctoral Symposium 4th IEEE International Symposium on Requirements Engineering.* IEEE Computer Society Press, Los Alamitos CA.

Kunda, D and Brooks, L (2000) Identifying and classifying processes (traditional and soft factors) that support COTS component selection: A case study. In: *Proceedings ECIS 2000.* University of Economics and Business Administration, Vienna, pp. 173–80.

Lam, W, McDermid, JA and Vickers, AJ (1997) Ten steps towards systematic requirements reuse. In: *Proceedings ISRE 97: 3rd IEEE International Symposium on Requirements Engineering*, Anapolis MD. IEEE Computer Society Press, Los Alamitos CA, pp. 6–15.

Larzelere, RJ and Huston, TL (1980) The Dyadic trust scale: Toward understanding the interpersonal trust in close relationships. *Journal of Marriage and the Family.*

Laurillard, D (1993) *Rethinking university teaching: A framework for the effective use of educational technology.* Routledge, London.

Layzell, PJ, Freeman, MJ and Benedusi, P (1995) Improving reverse-engineering through the use of multiple knowledge sources. *Journal of Software Maintenance: Research and Practice* 7, pp. 279–99.

Lehman, MM (1990) Uncertainty in computer applications. *Communications of the ACM* 33, pp. 584–86.

Lehman, MM and Ramil, JF (2000) Software evolution in the age of component based systems engineering. *IEE Proceedings: Software* 147, pp. 249–55.

Lenat, DB (1995) CYC: A large-scale investment in knowledge infrastructure. *Communications of the ACM* 38, pp. 33–38.

Leveson, N (1995) *Safeware: System safety and computers.* Addison Wesley, Reading MA.

Leveson, NG and Turner, CS (1993) An investigation of the Therac-25 accidents. *IEEE Computer* 26, pp. 18–41.

Lieberman, H (ed.) (2001) *Your wish is my command: Programming by example.* Morgan Kaufmann, San Francisco.

Liete, JC, St Anna, M and De Freitas, FG (1994) Draco-PUC: The technology assembly for domain oriented software development. In: *Frakes WBProceedings Third International Conference on Software Reuse*, Rio de Janeiro. IEEE Computer Society Press, Los Alamitos CA, pp. 94–100.

Lim, KY and Long, JL (1994) *The MUSE method for usability engineering.* Cambridge University Press, Cambridge.

Lohse, GL (2000) Usability and profits in the digital economy. In: McDonald, S, Waern, Y and Cockton, G (eds) *People and Computers XIV: Usability or Else; Proceedings BCS-HCI Conference.* Springer, Berlin, pp. 3–16.

Loucopoulos, P and Karakostas, V (1995) *System requirements engineering.* McGraw-Hill, London.

Lubars, M, Potts, C and Ritcher, C (1993) A review of the state of the practice in requirements modelling. In: Fickas, S and Finkelstein, ACW (eds) *Proceedings 1st International Symposium on Requirements Engineering.* IEEE Computer Society Press, Los Alamitos CA, pp. 2–14.

Luff, P and Heath, C (2000) The collaborative production of computer commands in command and control. *International Journal of Human–Computer Studies* 52, pp. 669–700.

Luff, P, Jirotka, M, Heath, C and Greatbatch, D (1993) Tasks and social interaction: The relevance of naturalistic analyses of conduct for requirements engineering. In: Fickas, S and Finkelstein, ACW (eds) *Proceedings 1st International Symposium on Requirements Engineering - RE93*, San Diego CA. IEEE Computer Society Press, Los Alamitos CA, pp. 187–90.

Macaulay, LA (1993) Requirements capture as a co-operative activity. In: Fickas, S and Finkelstein, ACW (eds) *Proceedings 1st International Symposium on Requirements Engineering.* IEEE Computer Society Press, Los Alamitos CA, pp. 174–81.

Macaulay, LA (1996) *Requirements engineering.* Springer-Verlag, Berlin.

MacLean, A, Young, RM, Bellotti, V and Moran, TP (1991) Questions, options and criteria: Elements of design space analysis. *Human–Computer Interaction* 6, pp. 201–50.

Maiden, NAM and Rugg, G (1994) Knowledge acquisition techniques for requirements engineering. In: *Proceedings Workshop on Requirements Elicitation for System Specification.* University of Keele, Keele.

Maiden, NAM and Sutcliffe, AG (1992) Exploiting reusable specification through analogy. *Communications of the ACM* 35, pp. 55–64.

Maiden, NAM and Sutcliffe, AG (1994) Requirements critiquing using domain abstractions. In: *Proceedings First International Conference on Requirements Engineering,* Colorado Springs CO. IEEE Computer Society Press, Los Alamitos CA, pp. 184–93.

Marchionini, G (1995) *Information seeking in electronic environments.* Cambridge University Press, Cambridge.

Maslow, AH, Frager, R, McReynolds, C, Cox, R and Fadiman, J (1987) *Motivation and personality* 3rd edn. Addison Wesley-Longman, New York.

Mazza, C, Fairclough, J, Melton, B, De Pablo, D, Scheffer, A and Stevens, R (1994) *Software engineering standards.* Prentice Hall, New York.

McCrae, RR and John, OP (1992) An introduction to the five factor model and its applications. *Journal of Personality* 60, pp. 175–215.

McDaniel, E, Olson, GM and Magee, JC (1996) Identifying and analyzing multiple threads in computer-mediated and face-to-face conversations. In: *Proceedings ACM 1996 Conference on Computer Supported Cooperative Work,* Boston MA. ACM Press, New York, pp. 39–47.

McMenamin, SM and Palmer JF (1984) *Essential systems analysis.* Yourdon Press, Englewood Cliffs NJ.

Milgram, S (1974) *Obedience to authority.* Tavistock, London.

Milner, R (1989) *Communication and concurrency.* Prentice-Hall, Hemel Hempstead.

Monk, AG and Wright, P (1993) *Improving your human-computer interface: A practical technique.* Prentice Hall, Hemel Hempstead.

Moscovici, S and Zavalloni, M (1969) The group as a polarizer of attitudes. *Journal of Personality and Social Psychology* 12, pp. 125–35.

Muller, MJ (1991) PICTIVE: An exploration in participatory design, user interface design process and evaluation. In: *Human Factors in Computing Systems: CHI 91 Conference Proceedings,* pp. 225–31.

Muller, MJ and Czerwinski, M (1999) Organising usability work to fit the full product range. *Communications of the ACM* 42, pp. 87–90.

Mylopoulos, J, Chung, L and Nixon, B (1992) Representing and using non-functional requirements: A process-oriented approach. *IEEE Transactions on Software Engineering* 18, pp. 483–97.

Mylopoulos, J, Chung, L and Yu, E (1999) From object-oriented to goal-oriented requirements analysis. *Communications of the ACM* 42, pp. 31–37.

Ncube, C and Maiden, NAM (1999) Guiding parallel requirements acquisition and COTS software selection. In: *Proceedings 4th IEEE International Symposium on Requirements Engineering,* Limerick, Ireland. IEEE Computer Society Press, Los Alamitos CA, pp. 133–40.

Neighbors, J (1984) The Draco approach to constructing software from reusable components. *IEEE Transactions on Software Engineering* 10, pp. 564–74.

Neighbors, J (1994) An assessment of reuse technology after ten years. In: Frakes, WB (ed.) *Proceedings 3rd International Conference on Software Reuse: Advances in Software Reusability.* IEEE Computer Society Press, Los Alamitos CA, pp. 6–13.

Nielsen, J (1993) *Usability engineering.* Academic Press, New York.

Norman, DA (1986) Cognitive engineering. In: Norman, DA and Draper, SW (eds) *User-centred system design: New perspectives on human–computer interaction.* Lawrence Erlbaum Associates, Hillsdale NJ.

Norman, DA (1988) *The psychology of everyday things.* Basic Books, New York.

Norman, DA (1999) *The invisible computer: Why good products can fail, the personal computer is so complex, and information appliances are the solution.* MIT Press, Cambridge MA.

Nunamaker, JF, Dennis, A, Valacich, J, Vogel, D and George, J (1991) Electronic meeting systems to support group work. *Communications of the ACM* 34, pp. 40–61.

Orth, M (1997) Interface to architecture: Integrating technology into the environment in the Brain Opera. In VanDerVeer, G, Henderson, A and Coles, S (eds) *Designing interactive systems: Proceedings DIS 97,* Amsterdam, pp. 265–76.

Papert, S (1980) *Mindstorms: Children, computers, and powerful ideas.* Basic Books, New York.

Pearl, J (1988) *Probabilistic reasoning in intelligent systems: Networks of plausible information.* Morgan Kaufmann, San Francisco.

Perrow, C (1984) *Normal accidents: Living with high-risk technology.* Basic Books, New York.

Pohl, K (1993) The three dimensions of requirements engineering. In: *Proceedings CAiSE 93, Paris.* Springer-Verlag, Berlin.

Pohl, K (1996) *Process-centred requirements engineering (Advanced software development, 5).* Research Studies Press, Taunton.

Porter, ME (1980) *Competitive strategy.* Free Press, New York.

Potts, C (1995) Invented requirements and imagined customers: Requirements for off-the-shelf software. Briefing for Working Group 2. In: *Proceedings Requirements Engineering 95,* IEEE Computer Society Press, Los Alamitos CA, pp. 128–30.

Potts, C (1999) ScenIC: A strategy for inquiry-driven requirements determination. In: *Proceedings 4th IEEE International Symposium on Requirements Engineering, Limerick, Ireland.* IEEE Computer Society Press, Los Alamitos CA, pp. 58–65.

Potts, C, Takahashi, K and Anton, AI (1994) Inquiry-based requirements analysis. *IEEE Software* 11, pp. 21–32.

Potts, C, Takahashi, K, Smith, J and Ora, K (1995) An evaluation of inquiry based requirements analysis for an Internet service. In: Harrison, MD and Zave, P (eds) *Proceedings 1995 IEEE International Symposium on Requirements Engineering (RE 95), York.* IEEE Computer Society Press, Los Alamitos CA, pp. 27–34.

Prieto-Diaz, R (1990) Domain analysis: An introduction. *Software Engineering Notes* 15, pp. 47–54.

Prieto-Diaz, R (1991) Implementing faceted classification for software reuse. *Communications of the ACM* 34, pp. 88–97.

Ramesh, B and Dhar, V (1992) Supporting systems development by capturing deliberations during requirements engineering. *IEEE Transactions on Software Engineering* 18, pp. 498–510.

Rasmussen, J (1986) *Information processing in human computer interaction: An approach to cognitive engineering.* North Holland, Amsterdam.

Rational Corporation (1999) *UML: Unified modelling language method.* http://www.rational.com (accessed 1999).

Reason, J (1990) *Human error.* Cambridge University Press, Cambridge.

Reason, J (1997) *Managing the risks of organizational accidents.* Ashgate, Aldershot.

Reeves, B and Nass, C (1996) *The media equation: How people treat computers television and new media like real people and places.* CLSI/Cambridge University Press, Stanford CA/Cambridge.

Repenning, A (1993) Agentsheets: A tool for building domain oriented-dynamic visual environments. Technical report, Dept of Computer Science, CU/CS/693/93. University of Colorado, Boulder CO.

Reubenstein, HB and Waters RC (1989) The requirements apprentice: An initial scenario. In: *Proceedings 5th International Workshop on Software Specification and Design, Pittsburgh.* IEEE Computer Society Press, Los Alamitos CA, pp. 211–18.

Robertson, J and Robertson, S (1999) *Mastering the requirements process.* Addison Wesley, Harlow.

Rockart, JF and Short, JE (1991) The networked organisation and the management of interdependence. In: Scott-Morton, MS (ed.) *The corporation in the 1990s: Information technology and organisational transfer.* Oxford University Press, New York.

Rolland, C, Achour, CB, Cauvet, C, Ralyte, J, Sutcliffe, AG and Maiden, NAM (1998) A proposal for a scenario classification framework. *Requirements Engineering* 3, pp. 23–47.

Roman, GC (1985) A taxonomy of current issues in requirements engineering. *Computer* 18, pp. 14–23.

Rosch, E (1985) Prototype classification and logical classification: The two systems. In: Scholnick, EK (ed.) *New trends in conceptual representation: Challenges to Piaget's Theory.* Lawrence Erlbaum Associates, Hillsdale NJ.

Ross, DT and Schoman, KE (1977) Structured analysis for requirements definition. *IEEE Transactions on Software Engineering* 3, pp. 6–15.

Rumbaugh, J (1991) *Object oriented modelling and design.* Prentice Hall, Englewood Cliffs NJ.

Ryan, M and Sutcliffe, AG (1998) Analysing requirements to inform design. In: Johnson, H, Nigay, L and Roast, C (eds) *People and Computers XIII; Proceedings BCS-HCI Conference, Sheffield.* Springer-Verlag, Berlin, pp. 139–57.

Sacks, H, Schegloff, EA and Jefferson, G (1974) A simple systematics for the organization of turn-taking in conversation. *Language* 50, pp. 696–735.

Schank, RC (1982) *Dynamic memory: A theory of reminding and learning in computers and people.* Cambridge University Press, Cambridge.

Scheer, AW (1994) *Enterprise-wide data modelling.* Springer-Verlag, Berlin.

Searle, JR (1969) *Speech acts.* Cambridge University Press, Cambridge.

Sheridan, TB (2000) Function allocation: Algorithm, alchemy or apostasy? *International Journal of Human-Computer Studies* 52, pp. 203–16.

Simon, HA (1973) The structure of ill-structured problems. *Artificial Intelligence* 4, pp. 181–201.

Singh, MG and Zeng, XJ (1996) Approximation properties of fuzzy systems generated by the min inference. *IEEE Transactions on Systems, Man and Cybernetics* 26, pp. 187–94.

Smart Procurement Implementation Team (2000) *The acquisition handbook: A guide to Smart Procurement* (3rd edn) Ministry of Defence, London.

Smith, DR (1992) Track assignment in an airtraffic control system: A rational reconstruction of system design. In: *Proceedings KBSE 92, Knowledge Based Software Engineering.* IEEE Computer Society Press, Los Alamitos CA, pp. 60–68.

Sommerville, I (1989) *Software engineering.* Addison Wesley, Wokingham.

Sommerville, I and Kotonya G (1998) *Requirements engineering: Processes and techniques.* John Wiley, Chichester.

Sommerville, I and Sawyer, P (1997) *Requirements engineering: A good practice guide.* John Wiley, Chichester.

Spivey, JM (1989) *The Z notation.* Prentice Hall, Englewood Cliffs NJ.

Sutcliffe, AG (1995a) *Human computer interface design* (2nd edn). Macmillan, London.

Sutcliffe, AG (1995b) Requirements rationales: Integrating approaches to requirements analysis. In: Olson, GM and Schuon, S (eds) *Proceedings DIS 95 Symposium on Designing Interactive Systems, Ann Arbor MI.* ACM Press, New York, pp. 33–42.

Sutcliffe, AG (1997) A technique combination approach to requirements engineering. In: *Proceedings ISRE 97: 3rd IEEE International Symposium on Requirements Engineering, Anapolis MD.* IEEE Computer Society Press, Los Alamitos CA, pp. 65–74.

Sutcliffe, AG (1998) Scenario-based requirements analysis. *Requirements Engineering* 3, pp. 48–65.

Sutcliffe, AG (1999a) Business modelling inter-process relationships. In: Bustard, D (ed.) *Proceedings of SMBPI Workshop on Systems Modelling for Business Process Improvement.* University of Ulster, Coleraine, pp. 185–204.

Sutcliffe, AG (1999b) A design method for effective information delivery in multimedia presentations. *New Review of Hypermedia and Multimedia, Applications and Research* 5, pp. 29–58.

Sutcliffe AG (1999c) User-centred design for multimedia applications. In: *Proceedings ICMCS 99, Vol. 1 IEEE Conference on Multimedia Computing and Systems, Florence.* IEEE Computer Society Press, Los Alamitos CA, pp. 116–23.

Sutcliffe, AG (2000) Requirements analysis for socio-technical system design. *Information Systems* 25, pp. 213–33.

Sutcliffe, AG (2001) Heuristic evaluation of website attractiveness and usability. GIST technical report G2001/1: Proceedings: 8th Workshop on Design Specification and Verification of Interactive Systems. Department of Computer Science University of Glasgow, Glasgow, pp. 188–99.

Sutcliffe, AG (2002) *The Domain Theory: Patterns for knowledge and software reuse.* Lawrence Erlbaum Associates, Mahwah NJ.

Sutcliffe, AG and Carroll, JM (1998) Generalizing claims and reuse of HCI knowledge. In: Johnson, H, Nigay, L and Roast, C (eds) *People and Computers XIII; Proceedings BCS-HCI Conference, Sheffield.* Springer-Verlag, Berlin, pp. 159–76.

Sutcliffe, AG and Carroll, JM (1999) Designing claims for reuse in interactive systems design. *International Journal of Human-Computer Studies* 50, pp. 213–41.

Sutcliffe, AG, Economou, A and Markis, P (1999) Tracing requirements errors to problems in the requirements engineering process. *Requirements Engineering* 4, pp. 134–51.

Sutcliffe, AG and Faraday, P (1994) Designing presentation in multimedia interfaces. In: Adelson, B, Dumais, S and Olson, J (eds) *CHI 94 Conference Proceedings Human Factors in Computing Systems Celebrating Interdependence, Boston MA.* ACM Press, New York, pp. 92–98.

Sutcliffe, AG and Lammont, N (2001) Business and IT requirements for B2B e-commerce. *New Product Development and Innovation Management*, pp. 353–70.

Sutcliffe, AG and Li, G (2000) Connecting business modelling to requirements engineering. In: Henderson, P (ed.) *Systems engineering for business process change: Collected papers from the EPSRC Research Programme.* Springer, London, pp. 91–105.

Sutcliffe, AG and Maiden, NAM (1990) How specification reuse can support requirements analysis. In: Hall, P (ed.) *Proceedings Software Engineering 90.* Cambridge University Press, Cambridge, pp. 489–509.

Sutcliffe, AG and Maiden, NAM (1992) Analysing the novice analyst: Cognitive models in software engineering. *International Journal of Man–Machine Studies* 36, pp. 719–40.

Sutcliffe, AG and Maiden, NAM (1993a) Bridging the requirements gap: Policies, goals and domains. In: *Proceedings 7th International Workshop of Software Specification and Design*, Redondo Beach CA. IEEE Computer Society Press, Los Alamitos CA, pp. 52–55.

Sutcliffe, AG and Maiden, NAM (1993b) Domain knowledge for requirements engineering. In: Rolland, C and Prakash, N (eds) *Proceedings IFIP WG 8.1 Conference on Information System Development Process*. North Holland, Amsterdam.

Sutcliffe, AG and Maiden, NAM (1998) The domain theory for requirements engineering. *IEEE Transactions on Software Engineering* 24, pp. 174–96.

Sutcliffe, AG, Maiden, NAM, Minocha, S and Manuel, D (1998) Supporting scenario-based requirements engineering. *IEEE Transactions on Software Engineering* 24, pp. 1072–88.

Sutcliffe, AG and Ryan, M (1998) Experience with SCRAM, a SCenario Requirements Analysis Method. In: *Proceedings IEEE International Symposium on Requirements Engineering: RE 98, Colorado Springs CO*. IEEE Computer Society Press, Los Alamitos CA, pp. 164–71.

Sutcliffe, AG, Ryan, M, Doubleday, A and Springett, MV (2000) Model mismatch analysis: Towards a deeper explanation of users usability problems. *Behaviour and Information Technology* 19, pp. 43–55.

Swain, AD and Weston, LM (1988) An approach to the diagnosis and mis-diagnosis of abnormal conditions in post-accident sequences in complex man machine systems. In: Goodstein, L, Andersen, H and Olson S (eds) *Tasks, errors and mental models*. Taylor and Francis, London.

Swartout, W and Balzer, R (1982) On the inevitable intertwining of specification and implementation. *Communications of the ACM* 25, pp. 435–30.

Thayer, R and Dorfman, M (1990) *System and software requirements engineering*. IEEE Computer Society Press, Los Alamitos CA.

Thimbleby, H (1990) *User interface design*. ACM/Addison Wesley, Reading MA.

Tracz, W (1995) *Confessions of a used program salesman: Institutionalizing software reuse*. Addison-Wesley, Reading MA.

Tseng, S and Fogg BJ (1999) Credibility and computing technology. *Communications of the ACM* 42, pp. 39–44.

Van Lamsweerde, A, Darimont, R and Massonet, P (1995) Goal directed elaboration of requirements for a meeting scheduler: Problems and lessons learnt. In: Harrison, MD and Zave, P (eds) *Proceedings 1995 IEEE International Symposium on Requirements Engineering*. IEEE Computer Society Press, Los Alamitos CA, pp. 194–203.

Van Lamsweerde, A and Letier, E (2000) Handling obstacles in goal-oriented requirements engineering. *IEEE Transactions on Software Engineering* 26, pp. 978–1005.

Verheijn, G and Van Bekkum, J (1982) NIAM: An information analysis method. In: Olle, TW, Sol, HG and Verrijn-Stewart, AA (eds) *Information system methodologies: A framework for understanding CRIS3 Task Group of IFIP WG8.1*. North-Holland, Amsterdam.

Williamson, OE (1981) The economics of organisations: The transaction cost approach. *American Journal of Sociology* 87, pp. 548–77.

Wilson, S, Bekker, M, Johnson, P and Johnson, H (1997) Helping and hindering user involvement: A tale of everyday design. In: Pemberton, S (ed.) *Proceedings Human Factors in Computing Systems CHI97, Atlanta GA*. ACM Press, New York, pp. 178–85.

Winograd, T and Flores, F (1986) *Understanding computers and cognition: A new foundation for design*. Addison Wesley, Reading MA.

Wright, P, Dearden, A and Fields, B (2000) Function allocation: A perspective from studies of work practice. *International Journal of Human–Computer Studies* 52, pp. 335–55.

Yu, E (1994) Modelling strategic relationships for process reengineering: Technical Report DKBS-TR-94-6. University of Toronto, Toronto.

Zave, P. (1995) Classification of research efforts in requirenments engineering. In: Harrison, MD and Zave, P (eds) *Proceedings 1995 IEEE International Symposium on Requirements Engineering*. IEEE Computer Society Press, Los Alamitos CA, pp. 214–16.

Index